MOLECULES AND MOLECULAR LASERS FOR ELECTRICAL ENGINEERS

Series in Electrical Engineering
Editor: S. A. Nasar

Bradley Molecules and Molecular Lasers for Electrical Engineers
de Jong AC Motor Design: Rotating Magnetic Fields in a Changing Environment

in preparation

Nasar and Boldea Electric Machines: A Modern Approach

MOLECULES AND MOLECULAR LASERS FOR ELECTRICAL ENGINEERS

Eugene B. Bradley

University of Kentucky
Lexington, Kentucky

○ HEMISPHERE PUBLISHING CORPORATION
A member of the Taylor & Francis Group

New York   Washington   Philadelphia   London

MOLECULES AND MOLECULAR LASERS FOR ELECTRICAL ENGINEERS

Copyright © 1990 by Hemisphere Publishing Corporation. All rights reserved. Printed in the United States of America. Except as permitted under the United States Copyright Act of 1976, no part of this publication may be reproduced or distributed in any form or by any means, or stored in a data base or retrieval system, without the prior written permission of the publisher.

1 2 3 4 5 6 7 8 9 0   B R B R   8 7 6 5 4 3 2 1 0 9

Cover design by Debra Eubanks Riffe.
A CIP catalog record for this book is available from the British Library.

Library of Congress Cataloging-in-Publication Data

Bradley, Eugene B.
   Molecules and molecular lasers for electrical engineers / Eugene B. Bradley.
      p.   cm. — (Series in electrical engineering)
   Includes bibliographical references.

   1. Molecular gas lasers. I. Title. II. Series.
TA1695.B73   1990
621.36'63—dc20
                                                                   89-26915
                                                                       CIP

ISBN 0-89116-788-9

CONTENTS

Foreword		ix
Preface		xi
1	**BASIC CONCEPTS**	**1**
	1 Introduction	1
	2 Spectroscopy	2
	3 Classification	7
2	**INTRODUCTION TO QUANTUM MECHANICS**	**9**
	1 Introduction	9
	2 The Schrodinger Equation	10
	3 The Time-Independent Schrodinger Equation	14
	4 Examples of Application to a Rotating Molecule and A Vibrating Molecule	14
	References	15
3	**THE RIGID PLANAR ROTATOR**	**17**
	1 Introduction	17
	2 The Fixed-Axis Rigid Rotor	18

	3	Absorption and Emission of Energy by the Rigid Rotor	21
	4	A Calculation From Elementary Physics	23
	5	The Use of Pure-Rotation Absorption as a Filter	23
	6	The Rotation Absorption Spectrum of a Precessing Diatomic Molecule	25
	7	Relative Intensities of Rotation Lines	26
	8	Centrifugal Stretching of the Internuclear Distance	28
	9	Example of Engineering Uses of Pure-Rotation Transitions in Molecules	28
		Problems—Pure Rotation	29
		References	29
4	A VIBRATING DIATOMIC MOLECULE		31
	1	Introduction	31
	2	Hooke's Law and the Effect of Anharmonicity	31
	3	Quantum Mechanical Treatment of a Heteronuclear Diatomic Oscillator	33
	4	Effects of Anharmonicity on Real Molecules	35
	5	Fermi Resonance	37
	6	Hot Bands	37
	7	Comparison of Molecular Resonances and Resonances of Discrete Circuits	39
5	VIBRATION—ROTATION		43
	1	Introduction	43
	2	Zero-Order Approximation	44
	3	The Vibration-Rotation Band	45
	4	Higher-Order Approximations	47
		Problems Vibration-Rotation	50
6	THE RAMAN EFFECT		51
	1	Introduction	51
	2	Classical Theory of the Raman Effect	52
	3	Quantum Theory of the Raman Effect	55
7	INTRODUCTION TO GROUP THEORY		59
	1	Reasons for Studying Methods of Group Theory	59
	2	Definition of An Operation	60
	3	Definition of a Group	60
	4	Examples of Groups	61

CONTENTS

	5	Other Definitions and Concepts	63
	6	Simple Isomorphism	65
	7	Order, Period	65
	8	Subgroup	65
	9	Conjugate Elements	66
	10	Classes of Elements	66
	11	Categories of Operations	66
	12	Coordinate Systems for Molecules	67
	13	The Water Molecule	70
	14	The Ethylene Molecule	72
	15	The Effect of Covering Operations on Coordinates	75
	16	Special Properties and Representations	80
	17	Useful Properties of Representations by Classes; The Character Table	83
	18	Summary	84
	19	Species	85
8	USES OF THE CHARACTER TABLE		87
	1	Introduction	87
	2	Transformations of Characters Under Sample Operations	88
	3	Calculation of the Characters of the Electric Dipole Moment for H_2O	89
	4	Infrared Selection Rules	90
	5	Character of the Polarizability	91
	6	Raman Selection Rules	95
	7	Characters of the Vibrational Coordinates	95
		References	107
9	LASER FUNDAMENTALS		109
10	MOLECULAR LASERS		117
	1	The CO_2 Laser	117
	2	Water Vapor Laser	122
		References	123
INDEX			125

FOREWORD

It is the purpose of this book to introduce electrical engineers to molecules in gas phase, some of their quantum properties and laboratory techniques used to study these properties, and concepts of molecular lasers. This book is not intended to be a laser textbook but rather serves to acquaint the reader with ideas behind modern engineering applications of molecular theory; the molecular laser being only one example. The group theory will prepare EE's to read advanced literature on (a) semiconductor energy states and (b) properties of adsorbed surface molecules, both of which are important topics in the IC industry.

PREFACE

"By the year 2012 the term electrical engineer will have disappeared from college catalogs to be replaced by 'electronics' or possibly 'electronic science'—perhaps the most challenging part of the education of the electronic scientist of 2012 will be associated with energy at wavelengths extending from millimeters into fractions of millimeters to the infrared, visible and ultraviolet regions"

                              Frederick E. Terman
                              "Education in 2012 for
                              Communication and Electronics"
                              Proceedings of the IRE
                              Fiftieth Anniversary Issue
                              May, 1962

    The fact that molecules and atoms may be used to perform electrical circuit functions seems unlikely to some electrical engineers, but let me assure you that it is entirely possible! The rapid growth of quantum electronic devices, thin-film microwave techniques, and optical memory systems leaves little doubt that electrical engineers are becoming more familiar with applications of the atomic and molecular behavior of materials in the solid, liquid, or gas phase. Since 1962, a wide variety of molecular and atomic systems have been designed to perform circuit functions at millimeter, submillimeter, and visible wavelengths. At these wavelengths, molecules and atoms are used to amplify, amplify parametrically, detect, generate radiation, modulate, demodulate, measure radiation power, and filter. From knowledge of atomic and molecular behavior there have arisen new practical systems such as lasers (atomic and molecular), laser communications systems, remote sensing systems, optical memories for computers, etc. The impact of such systems upon the communications area alone is tremendous. Matchbox-size computer memories are a reality, and the read in-read out time is extremely fast.

Just a few short years ago, the infrared, visible, and ultraviolet regions of the electromagnetic spectrum were almost exclusively the province of physicists and chemists. Today, some electrical engineers know about the quantum behavior of materials at wavelengths less than one millimeter, and they know how such behavior may be put to practical use in devices and systems. Electrical engineers who design quantum electronic devices and systems really work in a multidisciplinary field, drawing upon knowledge of atomic and molecular spectroscopy, quantum theory, solid-state physics, various areas of chemistry, optics, etc. Engineers may acquire knowledge of the field from several approaches, each of which is suitable for a specific set of goals, but almost always they use basic principles of molecular and atomic behavior, optics, solid-state theory, and quantum theory as a core of knowledge from which to diverge to new and special areas.

This text presents basic molecular spectral phenomena (absorption and emission) which occur in the infrared and visible regions of the electromagnetic spectrum. It is from the theoretical interpretation and the practical use of these spectra that some new materials and new quantum electronic devices and systems have evolved. A little reflection upon the history of electrical engineering should convince the reader that some big breakthroughs in the discipline are rooted in *materials*. Certainly the recent discovery of high T_c superconductors is another big breakthrough. The electronics of the year 2012 could be here today if materials research could be advanced more rapidly.

CHAPTER

1

BASIC CONCEPTS

1 INTRODUCTION

We begin our study of molecules and molecular gas lasers with an introduction to molecules and the techniques that are used to study them. It is assumed that the reader has some ideas about molecular structure from elementary physics and chemistry. Each atom of a molecule consists of a positively charged nucleus surrounded by electrons. Inside the molecular framework the nuclei occupy nearly fixed relative positions and the electrons, which are much less massive than the nuclei, move about inside the framework. The interatomic forces which bind the atoms into a molecule are often represented by small springs. We shall consider only the properties of isolated molecules in gases, treating the molecule as a three-dimensional structure consisting of the nuclear masses joined together by small springs and ignoring the electronic masses.

The equilibrium configuration of the nuclei is related to the state of motion of the associated electrons inside the framework, and various electronic energy states are possible, each corresponding to a different minimum potential energy of the molecule and each going with a different equilibrium configuration of the nuclei. At first we shall be concerned only with the geometrical structure of a molecule and how the vibration and rotation of that structure influences the absorption or emission of microwave and infrared energy. We shall see later in Chapter 6 on the Raman effect just how the motions of the nuclear masses affect the electronic energy states of the molecule in such a way that the molecule will scatter visible light. The light frequencies are shifted both above and below the incident frequency (the Raman effect)—an optical "analog" of amplitude modulation where upper and lower sideband frequencies occur.

Absorption, emission and scattering of electromagnetic energy by molecules in the three states of matter are indeed "fingerprints" of the particular molecular structure which interacts with electromagnetic radiation. The *spectra*, which are records of electromagnetic interactions with molecules, are unique for a given molecular structure. An absorption spectrum usually is some plot of frequency (abscissa) *vs* percent transmission (ordinate) or of some units which are proportional to these. An emission spectrum has its ordinate-abscissa scaled in relative intensity *vs* frequency. A vast amount of useful physical information may be obtained from these "prints" if one knows how to interpret them, information which may be put to practical use. For example, electrical engineers might use the information to design, *lasers, filters, modulators*, etc. Chemists, physicists and biochemists often study molecular structure and identify unknown compounds, astronomers study the molecular structure of stars, narcotics agents detect the presence of illegal drugs, etc.

The frequency range encompassed by characteristic molecular motions are listed in Table 1, along with the physical information available from the spectra which document such motions and their interactions with electromagnetic radiation. *The particular type of molecular motion may be exploited to study the physical properties of the molecule itself—or the interaction of electromagnetic radiation with molecular motion may be exploited to engineer a useful device such as a molecular laser.*

Let us see how the frequencies in Table 1 fit into the electromagnetic spectrum as an electrical engineer might view it. Displayed in Fig. 1.1 is the electromagnetic spectrum divided into regions according to frequency and wavelength. Typical "circuit elements" are listed for each region. At frequencies higher than 300 GHz "circuits" do not look like their counterparts at lower frequencies, and quasi-optical devices or optical elements are used because the wavelengths are very short. In the 1960s the electromagnetic spectrum had not really been tapped: the services extended mainly from 20 kHz to 40,000 MHz (about 6 decades) and beyond lay vast reaches in the frequency domain as yet unexplored. In the regions above 300 GHz there are many decades available for communications systems, and information bandwidths are thousands of times larger than at lower frequencies.

2 SPECTROSCOPY

The absorption or emission of infrared energy by molecular motion (vibration or vibration-rotation) is studied experimentally by *infrared spectroscopy*, and the scattering of light by molecules is studied by *Raman spectroscopy*. The term *spectroscopy* is used widely in science to mean detection, resolution, and recording of energetic phenomena which occur in nuclei, atoms, or molecules. A *spectrometer* is often used to make an instantaneous observation and a *spectrograph* makes some type of permanent record. A *spectrophotometer* ratios optically two light beams to cancel out

Table 1. Characteristic Molecular Motions and Frequency Ranges of Their Occurrence

Spectral Region	Type of Molecular Motion	Frequency (sec^{-1})	Typical Information Obtained from Spectra
Microwave	Pure rotation of heavy molecules	$10^9 - 10^{11}$	Dipole moments; interatomic distances; dispersion; nuclear interactions
Far infrared	Pure rotation of light molecules; vibration of heavy molecules	$10^{11} - 10^{13}$	Interatomic distances; force constants of bonds; structure; dispersion.
Infrared	Vibration Vibration-rotation	$10^{13} - 10^{14}$	Interatomic distances; molecular charge distribution; force constants of bonds; structure; thermodynamic properties; dispersion.
Raman Scattering of Visible Light	Vibrations, rotations Vibration-rotation	$10^{11} - 10^{14}$	Interatomic distnaces, force constants of bonds; molecular charge distributions; structure.
Visible; ultraviolet	Electronic transitions	$10^{14} - 10^{16}$	Properties listed above; dissociation energies.

any undesired background. Typically, electromagnetic energy from a wideband source interacts with a molecular sample (solid, liquid or gas) and the light which is absorbed, emitted, or scattered by the sample is frequency analyzed by a spectrometer or a spectrograph. The basic features of such a spectrometer are shown in Fig. 1.2. The spectrum of the light (be it infrared or visible) which is seen by the detector includes (a) the frequencies (called spectral lines) which are absorbed, emitted, or scattered by the sample, and (b) the relative intensities and shapes of these lines. Ideally a spectral line is the monochromatic image of a slit, but actually the slit passes a narrow band of frequencies because it has a finite aperture. Thus a narrow band of frequencies is actually recorded, rather than a single frequency. Spectral lines and their intensities may be interpreted to yield useful physical information.

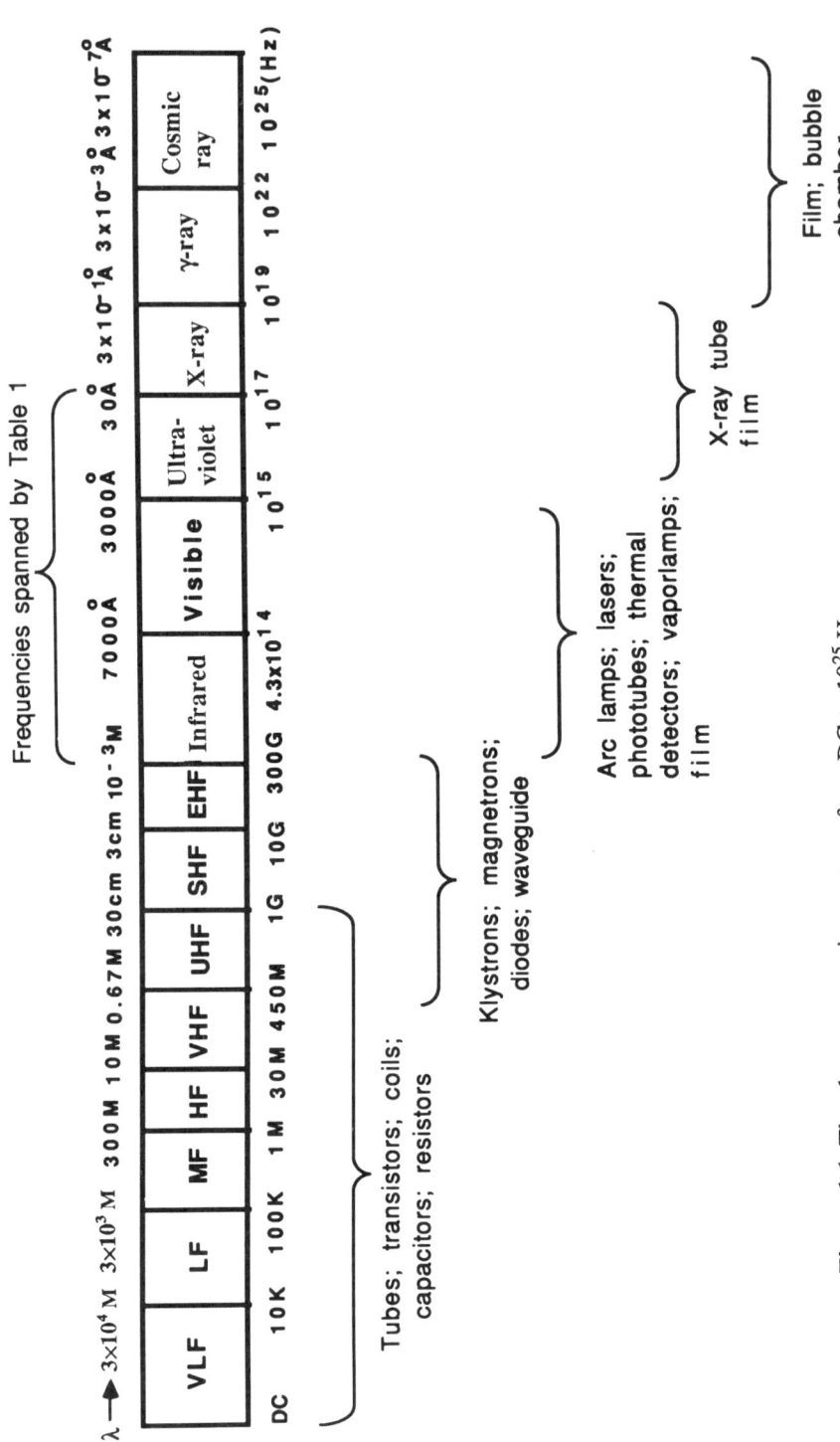

Figure 1.1 The electromagnetic spectrum from DC to 10^{25} Hz.

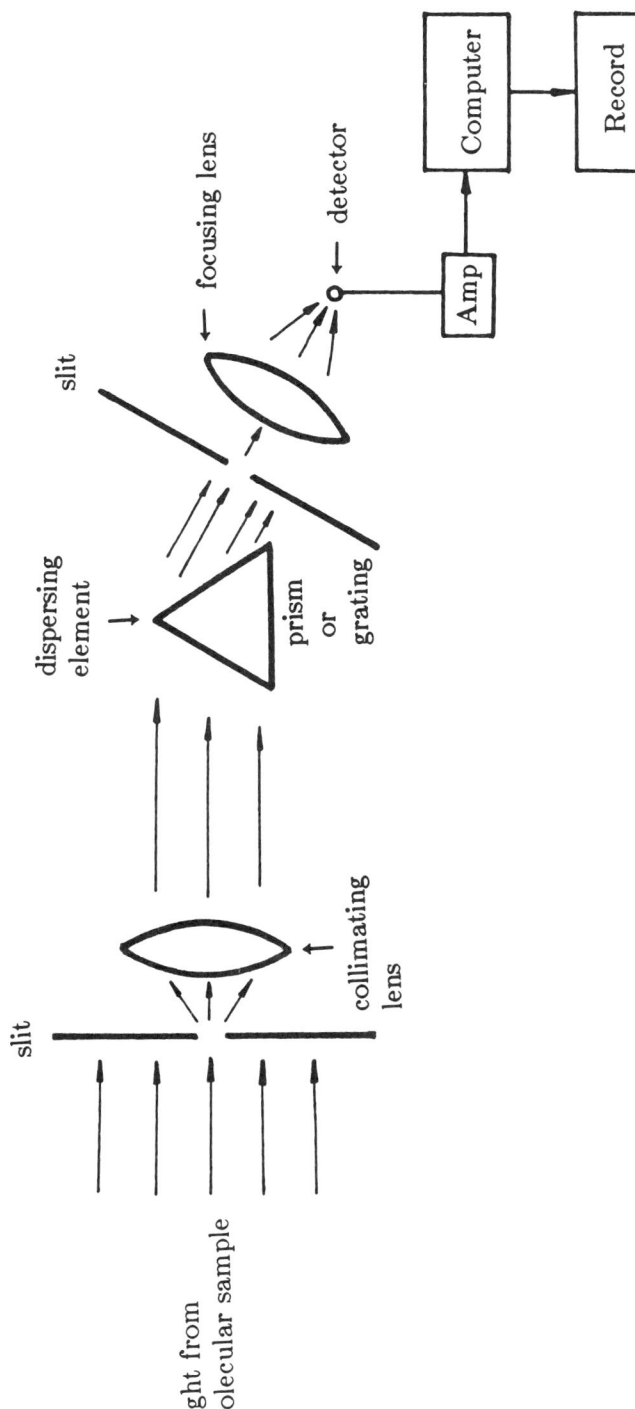

Figure 1.2 A dispersive spectrometer.

To obtain higher resolution, it is necessary to decrease the widths of the slits, resulting in a smaller fraction of energy at the detector. For example, halving the slit width produces twice the resolution, but the scan speed (number of frequency units scanned per second) must be one-fourth of the original scan speed to yield the original signal-noise ratio.

A totally different approach to produce an infrared spectrum of a sample is the use of a Fourier transform spectrometer. This spectrometer is an interferometer, coupled with the necessary computer and software. Instead of recording a narrow band of frequencies at a time, all frequencies of the spectral range of interest are observed simultaneously. The signal-noise enhancement realized by simultaneous observation of the entire spectral range is at least a factor of 5 better.

The conventional dispersive spectrometer requires large radii for its collimation mirrors to obtain high resolution which requires small solid angles. (High resolution necessitates large focal length-to-diameter ratio). However, an interferometer collects energy at large angles with almost no limit on resolution. The interferometer has very high resolving power, excellent frequency accuracy and fast scan time. For a given scan time, the interferometer will have higher resolution and a higher signal-to-noise ratio than a dispersive instrument.

The spectral range of an interferometer is determined by the thickness and refractive index of the beam splitter. The source signal is divided by the beam splitter into two electromagnetic waves traveling perpendicular to one another (paths 1 and 2). After reflection from M1 and M2, the waves are recombined at the beam splitter. The mirror M1 is movable, and it varies the optical path length along path 1. At zero optical path difference, all frequencies in the two waves will add constructively. At nonzero path difference at least some destructive interference will occur. The signal seen by the detector is then one of varying intensity I(x), where x is optical path difference. The Fourier transform of I(x) is the frequency spectrum of the sample.

Spectroscopists usually do not work with "frequency" units as electrical engineers know it, but rather in units of reciprocal wavelength. Wavelengths are measured in Angstrom units (Å), micrometers (μm), or millimicrons (mμ) where

$$1\text{Å} = 10^{-8} cm\,;\, 1\,\mu m = 10^{-6}\, meter = 10^4\,\text{Å}.$$

It is common practice to state reciprocal wavelength in "wavenumbers" or number of waves per unit length, $1/\lambda$. The frequency ν of a spectral line, is related to its wavenumber σ by

$$\sigma = 1/\lambda = \nu/c \ (cm^{-1}) \tag{1.1}$$

where c is the speed of light in vacuum. As an example, the fundamental vibrational frequency of the hydrogen chloride molecule (HCl) is 3000 cm^{-1} or 3×10^{13} cycles per second.

Frequency and energy are related. The energy carried by one photon is $E = h\nu$ where h is Planck's constant. Wavenumber is written as $\sigma \equiv E/hc \ (cm^{-1})$; thus wavenumber is used to state both frequency and energy, although it is really neither but only proportional to them. The quantity E/hc is referred to as a *term* value.

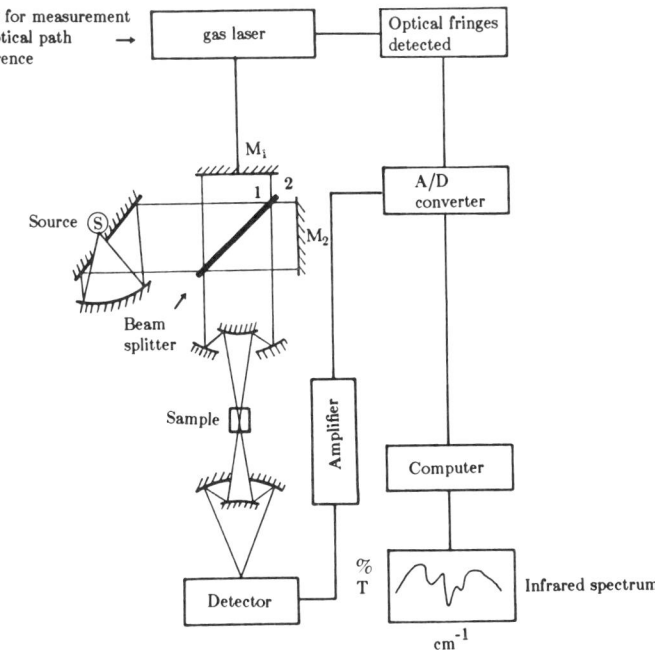

Figure 1.3 Block diagram of a Fourier transform spectrometer.

3 CLASSIFICATION

Molecules are classified in several ways, each one of which is useful. They are classified first according to the number, N, of constituent atoms:

(a)	monatomic	$N = 1$.	Example:	H (hydrogen)
(b)	diatomic	$N = 2$.	Example:	HCl (hydrogen chloride)
(c)	triatomic	$N = 3$.	Example:	CO_2 (carbon dioxide)
(d)	tetratomic	$N = 4$.	Example:	NH_3 (ammonia)
(e)	pentatomic	$N = 5$.	Example:	CCl_4 (carbon tetrachloride)
(f)	polyatomic	$N > 2$.		

Classification according to geometrical shape

(a)	point mass.	Example:	H (hydrogen)
(b)	linear.	Example:	CO_2 (carbon dioxide)
(c)	planar.	Example:	C_6H_6 (benzene)
(d)	non-planar.	Example:	NH_3 (ammonia)

Classification according to symmetry as a rigid rotating structure.

In classical mechanics every material body, regardless of shape, possesses three mutually perpendicular axes of rotation about which the body can be spun without wobbling (the principal axes). (No wobble means that the torque vector and the angular momentum vector point in the same direction). The moment of inertia about any of these three axes is called a principal moment.

(a) Asymetrical—all three principal moments of inertia are different.
(b) Symmetrical—two principal moments of inertia are equal, but the third principal moment is different.
(c) Spherical—the three principal moments of inertia are equal.
Classification according to chemical terminology.

Problem: One of the fundamental modes of vibration of the carbon dioxide molecule, CO_2, is observed at 668 cm^{-1}.
   (a) Compute the corresponding wavelengths in Å and μm.
   (b) Compute the frequency of the vibration in cycles per second.

CHAPTER

2

INTRODUCTION TO QUANTUM MECHANICS

1 INTRODUCTION

Some electrical engineers may not have a background in quantum mechanics or in modern physics, but that is not necessary to understand what follows.

The existence of discrete, band, or continuous spectra which occur in the electromagnetic spectrum cannot be explained adequately by classical mechanics. Atoms and molecules do not behave as our classical senses and experiences tell us they should, i.e., not just any energy is allowed to be absorbed or emitted by atoms or molecules. Quantum mechanics must be employed to calculate and predict atomic and molecular behavior in the presence of electromagnetic fields. In addition to spectra, other experimental evidence abounds to convince us that the world is indeed a quantized one and furthermore, that particles do behave like waves and vice versa. It is suggested that you read about some of the classic experiments which demonstrate conclusively the quantized wave-particle duality of matter and of electromagnetic energy—experiments such as the double-slit experiment,[1] the photoelectric effect,[2] Compton scattering,[3] and electron diffraction.[4]

The beginnings of the quantum theory of matter occurred early in the 20th century when Planck and Bohr advanced revolutionary ideas to explain (correctly) observed phenomena which classical mechanics had failed to predict. Planck[5] explained the energy distribution with frequency of black body radiation by assuming the atomic oscillators in the body are quantized according to $E = h\nu$, where h is Planck's constant, a universal physical constant. The quantity $h\nu$ is called a photon. Bohr,[6] in 1914, laid the foundation for the correct interpretation of the spectra of atoms and molecules with these postulates:

(a) An atom may exist in stable energy states without radiating electromagnetic energy.
(b) Absorption or emission of electromagnetic energy occurs when an atomic system changes from one energy state to another.
(c) The emission or absorption process corresponds to a photon of radiant energy $h\nu = hc\sigma = E' - E''$, where $E' - E''$ is the difference in energy between states of the atomic system.

Louis de Broglie[7] hypothesized that the momentum, p, of a photon is related to its wavelength by $\lambda = h/p$, which may also be written as $\lambda = 2\pi/k$ or $p = \hbar k$, where $\hbar = h/2\pi$.

Bohr also stated a *correspondence principle*,[8] *viz.*, that quantum mechanics applies to atomic systems and classical mechanics applies to macroscopic systems, but in some intermediate range the two types of mechanics should predict the same results. By a limiting process in which an atomic system approaches the macroscopic system, one may often employ classical mechanics to calculate important aspects of atomic or molecular behavior. We shall use this correspondence principle later on to determine selection rules governing transitions between energy states in molecules, but first we shall see how quantum mechanics is used to determine energy states of molecules. *It is from knowledge of these energy states that many useful engineering applications arise.*

2 THE SCHRODINGER EQUATION

Quantum mechanics is based upon the assumption that one may only calculate *probabilities* of obtaining certain values for the observables of a system, observables such as energy, momentum, position, angular momentum, etc. The energy of a stable atomic system has discrete values only.

The interrelationship between matter and energy (wave-particle duality) is described in a probabilistic way by the Schrodinger equation. The equation is indeed a "wave equation" but the wave is a wave of probability whose absolute square has a physical interpretation. The waves of probability are analogous to light waves because for light we can predict the probability of finding a photon at a certain point in space. The light intensity is proportional to the square of the absolute value of the wave amplitude; the probability for finding an atomic-size particle at a point in space is proportional to the absolute square of the wave function.

The Schrodinger equation is not really derivable in the strict sense of the word, just as Newton's second law of motion is not really derivable either. The equation's existence is attributable to the genius of Schrodinger, but it can be "derived" by analogy with the classical wave equation. We shall do this now.

The quantum mechanical wave equation must satisfy the following conditions:
(a) No preferred directions (invariant under rotation of axes).
(b) Does not contain time-changing parameters.

The familiar classical wave equation in one dimension is

$$\nabla^2 y = \frac{1}{v^2} \frac{\partial^2 y}{\partial t^2} \tag{2.1}$$

where y is the wave amplitude, v is the phase velocity of the wave, and t is time.

We examine the suitability of this equation to describe a wave of probability. Let a wave of probability $\psi(\underline{r},t)$ be a function of space and time. The vector $\underline{r}$ has three components and ψ may also be a complex function. The simplest wave to describe is a plane wave so we write

$$\psi(\underline{r},t) = e^{j(\underline{k}\cdot\underline{r}-\omega t)}, \tag{2.2}$$

a plane wave in 3d-space which propagates in the direction $\underline{k}$. If the wave function ψ is substituted in the wave equation for the wave amplitude y, then

$$\nabla^2 \psi = \nabla \cdot (\nabla \psi) = \nabla \cdot \underline{k}\, e^{j(\underline{k}\cdot\underline{r}-\omega t+\pi/2)} = -k^2 \psi = \frac{-p^2}{\hbar^2} \psi \tag{2.3}$$

and

$$\frac{\partial^2 \psi}{\partial t^2} = \frac{\partial}{\partial t}(-j\omega\psi) = \omega^2 \psi = \frac{E^2}{\hbar^2} \psi \tag{2.4}$$

Write $1/v^2$ as K, then

$$\nabla^2 \psi = K \frac{\partial^2 \psi}{\partial t^2} \tag{2.5}$$

or

$$\frac{-p^2}{\hbar^2} \psi = K \frac{E^2}{\hbar^2} \psi \tag{2.6}$$

and

$$-p^2 = KE^2 \tag{2.7}$$

which is not true non-relativistically, because $p^2/2m = E$ for a free particle. Thus

$$E^2 = p^4/4m^2 = -p^2/K, \tag{2.8}$$

where

$$K = -4m^2/p^2 \tag{2.9}$$

and the parameter K is a function of momentum in contradiction to condition (b).

Therefore let us try the equation shown below with a first partial derivative with respect to time.

$$\nabla^2 \psi = K \frac{\partial \psi}{\partial t} \tag{2.10}$$

or

$$-\frac{p^2}{\hbar^2} = \frac{KjE}{h}, \text{ where } E = p^2/2m \tag{2.11}$$

Substituting for E,

$$\frac{p^2}{\hbar^2} = -\frac{jKp^2}{2mh} \tag{2.12}$$

or

$$K = \frac{-2jm}{\hbar} \tag{2.13}$$

so

$$\nabla^2 \psi = \frac{-2mj}{\hbar} \frac{\partial \psi}{\partial t} \tag{2.14}$$

or

$$-\frac{\hbar^2}{2m} \nabla^2 \psi = j\hbar \frac{\partial \psi}{\partial t} \tag{2.15}$$

which is Schrodinger's wave equation. Notice that this equation is satisfied by a plane wave of the form $e^{j(k \cdot r - \omega t)}$. The wave function is found by solving this partial differential equation for ψ using the appropriate boundary conditions. The wave function must be finite, single-valued and continuous. For any values of r and t,

$$|\psi|^2 d\tau \sim P(r,t)\, d\tau \tag{2.16}$$

where $P(r,t)$ is the probability that the particle is in the element of volume $d\tau$. The proportionality means that

$$\frac{|\psi(r_1,t_1)|^2 d\tau_1}{|\psi(r_2,t_2)|^2 d\tau_2} = \frac{P(r_1,t_1)\, d\tau_1}{P(r_2,t_2)\, d\tau_2}. \tag{2.17}$$

The quantity $|\psi(r,t)|^2$ is the probability density, (the probability of finding the particle in volume $d\tau$) if $\psi(r,t)$ is a normalized function,

$$\int_{\text{all space}} |\psi|^2 d\tau = 1 \tag{2.18}$$

which means that if we look throughout all space we shall find the particle.

As an example of the normalization of ψ, suppose that

$$\int_\infty |\psi|^2 d\tau = A \geq 0 \ (A = 0 \text{ only if } \psi = 0) \tag{2.19}$$

then if $\psi' = \frac{\psi}{\sqrt{A}}$

$$\int_{\infty} |\psi'|^2 \, d\tau = 1. \tag{2.20}$$

A different form of the Schrodinger equation results when the plane wave is written in the form

$$\psi = e^{j(\underline{p}\cdot\underline{r}/\hbar - Et/\hbar)} \tag{2.21}$$

In this expression for ψ, note that $\underline{k} = \underline{p}/\hbar$ and $\omega = E/\hbar$. We calculate the left side and right side separately, then combine the results.

$$(L.S.) \quad \frac{-\hbar^2}{2m}\nabla^2\psi = \frac{-\hbar^2}{2m}\left[\frac{-p^2}{\hbar^2}\psi\right] = \frac{p^2}{2m}\psi \tag{2.22}$$

$$(R.S.) \quad \frac{\partial \psi}{\partial t} = \frac{-jE\,\psi}{\hbar} \tag{2.23}$$

or

$$E\psi = j\hbar \frac{\partial \psi}{\partial t} \tag{2.24}$$

We equate (2.22) and (2.24), so

$$\frac{p^2}{2m}\psi = E\psi \tag{2.25}$$

or

$$H\psi = E\psi \tag{2.26}$$

where $H = \dfrac{p^2}{2m}$, the total energy of the particle (if the potential energy is defined as zero).

In general, $H = \dfrac{p^2}{2m} + V(\underline{r})$, where H is the Hamiltonian of classical mechanics, viz., the sum of kinetic energy and potential energy of a "system". Thus, equation (2.26) may be written as

$$\left[\frac{p^2}{2m} + V(\underline{r})\right]\psi = E\psi \tag{2.27}$$

where by comparison with equation (2.22) the term $\dfrac{p^2}{2m}$ is interpreted as an *operator*, i.e., $-\dfrac{\hbar^2}{2m}\nabla^2$.

Thus from equation (2.26), if the total energy of the atomic system is written down and then treated as an operator on the wave function, this operation produces the allowed energies E of the system.

14 MOLECULES AND MOLECULAR LASERS FOR ELECTRICAL ENGINEERS

3 THE TIME-INDEPENDENT SCHRODINGER EQUATION

If the operator H is not an explicit function of time, then another important useful form of (2.26) may be obtained. The plane wave may be written as

$$\psi = \left[e^{i\underline{k}\cdot\underline{r}}\right] \cdot \left[e^{-j\omega t}\right] = U_1(\underline{r}) \cdot U_2(t). \tag{2.28}$$

If this form of ψ is put into equation (2.26) then we get

$$H\;[U_1(\underline{r}) \cdot U_2(t)] = E\;U_1(\underline{r}) \cdot U_2(t). \tag{2.29}$$

H will contain partial derivatives, as we have seen, so

$$[H\;U_1(\underline{r})] \cdot U_2(t) + U_1(\underline{r})\;[HU_2(t)] = E\;U_1(\underline{r}) \cdot U_2(t). \tag{2.30}$$

Now if $H \neq H(t)$ explicitly, then the second term on the left side of Eq. (2.30) is zero, and

$$H\;U_1(\underline{r}) = E\;U_1(\underline{r}). \tag{2.31}$$

This last equation is no longer the Schrodinger equation we saw before but it is often referred to as the time-independent Schrodinger equation. It is also called an eigenfunction—eigenvalue equation. Keep in mind that H is an operator, so it isn't legal to cancel $U_1(\underline{r})$ on both sides of equation (2.31). The equation (2.31) says that an operator H operates on a function $U_1(\underline{r})$, the eigenfunction, to prodce a *real number E*, the eigenvalue, times the same function.

If we know the total classical energy of an atomic system, i.e., the Hamiltonian which is the potential energy plus kinetic energy, then *this total energy is treated as a mathematical operator* which operates upon a function $U_1(\underline{r})$ to produce the energy level E of the system. If any wave function has the form of (2.28), then the probability density is not a function of the time, and the system is said to be in a "stationary state". Equation (2.31) is an extremely useful one, and we shall use it to perform several important calculations concerning the molecular energies of vibration and rotation.

4 EXAMPLES OF APPLICATION TO A ROTATING MOLECULE AND A VIBRATING MOLECULE

In Chapters 3 and 4 you will see examples of the application of the quantum theory to (a) a rigid planar molecule which rotates about its center of mass, and (b) a diatomic molecule which vibrates but does not rotate. The applications will require the construction of the Hamiltonian for (a) and for (b). Then the time-independent Schrodinger equation will be solved for the energy levels which are permitted by quantum theory. Molecules absorb or emit electromagnetic energy by *changes* in energy levels, but only certain changes are permitted by the *selection rules*. These rules are explained and applied to energy levels of rotating and vibrating molecules.

REFERENCES

1. *Principles of Modern Physics*, Robert B. Leighton, McGraw-Hill, New York, 1959, p. 81.
2. *Ibid.*, p. 67.
3. *Ibid.*, p. 432.
4. *Ibid.*, p. 83.
5. *Ibid.*, p. 64.
6. *Ibid.*, p. 72.
7. *Ibid.*, p. 81.
8. *Ibid.*, p. 88.

CHAPTER

3

THE RIGID PLANAR ROTATOR

1 INTRODUCTION

We begin the study of molecular energies with the energies of rotation of a molecule about its center of mass. The simplest example is a diatomic molecule treated as a rigid planar rotator, and we will examine this model in detail. The spectral pattern is relatively simple for this case but the spectral pattern becomes more complex for polyatomic molecules. It is convenient to classify molecules under three previously mentioned categories in order to study their rotational energy states. These categories are:
(a) Spherical top—the three principal moments of inertia are equal.
(b) Symmetric top—two principal moments of inertia are equal.
(c) Asymmetric top— the three principal moments of inertia are different.
Each category has its own particular energy scheme. There is also the special case of a symmetric top (linear polyatomic molecules) which has one principal moment of inertia zero (or very small) and the other two principal moments of inertia equal.

A molecule exhibits a pure-rotation spectrum only if it has a permanent electric dipole moment. Whether it absorbs electromagnetic energy in the far infrared or in the microwave region of the spectrum depends upon the magnitude of that permanent electric dipole moment and upon the value of the moment of inertia of the molecule. Most polyatomic molecules exhibit transitions in the microwave while light diatomics such as the hydrogen halides (HF, HCl, HBr, HI) exhibit some transitions in the far infrared. There are also instances where, under high pressure, molecules which normally do not exhibit pure-rotation transitions (no permanent electric dipole) can be made to absorb far infrared or microwave energy. This happens because under high pressure molecular collisions induce temporary electric dipoles.

Earlier it was stated that molecules are modeled as mass-spring systems. Consider such a model of a diatomic molecule which is rotating about its center of mass. As the speed of rotation increases, the spring stretches from centripetal force and the original moment of inertia of the molecule changes. The problem is a tractable one in classical mechanics and any energy of rotation is allowed as long as the molecule holds together. However, not just any rotational energy is permitted for a molecule; unlike its classical counterpart the rotational energies of a molecule are quantized. When a molecule rotates, it may absorb or emit microwave or far-infrared energy, and some far-infrared gas lasers such as water or sulfur dioxide use pure-rotation energy changes to produce coherent radiation in the far infrared. Examples will be discussed later in Chapter 10.

2 THE FIXED-AXIS RIGID ROTOR

Consider a diatomic or a linear molecule with a stiff spring between the atoms so that the molecule does not stretch as it rotates; such a model is called a *rigid* rotor. The axis of rotation does not change its orientation in space, i.e., it does not precess. The model is shown in Fig. 3.1 below. The axis of the coordinate system is taken at the center of mass of the molecule, and the axis of rotation is perpendicular to both x and y. For example, this model could represent HCl (hydrogen chloride) a diatomic molecule. This diatomic rotator may also be represented as a one-particle rotator by finding the reduced mass of the molecule and the radius of gyration, R, of the molecule. The moment of inertia is

$$I = \sum_i m_i r_i^2 = m_1 r_1^2 + m_2 r_2^2 \tag{3.1}$$

taken about an axis perpendicular to the x-y plane and passing through the center of mass. From the definition of the center of mass

$$m_1 r_1 = m_2 r_2 \tag{3.2}$$

or

$$= m_2 (R - r_1). \tag{3.3}$$

Solve for r_1

$$r_1 = \frac{m_2}{m_1 + m_2} R .$$

Substitute in equation (3.2) for $r_1 = R - r_2$ and

THE RIGID PLANAR ROTATOR   19

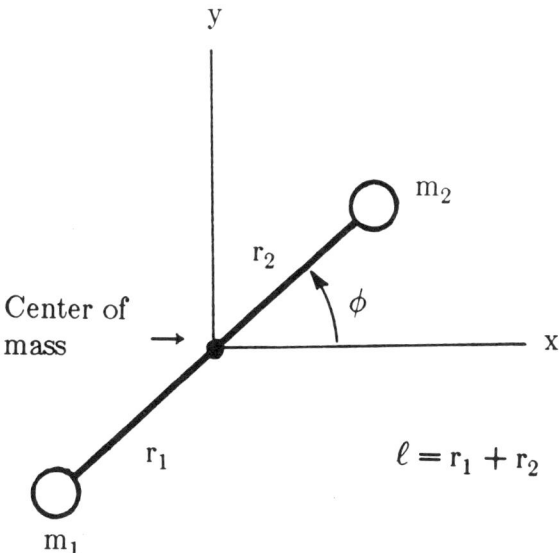

Figure 3.1 A rigid planar rotator.

$$r_2 = \frac{m_1}{m_1 + m_2} R \quad (3.4)$$

Substitute equation (3.4) and (3.3) into equation (3.1).

$$I = m_1 \left[\frac{m_2 R}{m_1 + m_2}\right]^2 + m_2 \left[\frac{m_1 R}{m_1 + m_2}\right]^2$$

$$= \frac{\left[\frac{m_1 m_2 R}{m_1 + m_2}\right]^2}{m_1} + \frac{\left[\frac{m_1 m_2 R}{m_1 + m_2}\right]^2}{m_2}$$

$$= \frac{\mu^2 R^2}{m_1} + \frac{\mu^2 R^2}{m_2} \quad (3.5)$$

where $\mu = m_1 m_2/(m_1 + m_2)$, the reduced mass.

$$I = \mu^2 R^2 \left[\frac{1}{\mu}\right] = \mu R^2 \quad (3.6)$$

This equation tells us that we may place a mass μ at a distance R (radius of gyration) from the axis of rotation and produce the same moment of inertia as masses m_1 and m_2 produce at distances r_1 and r_2 from the axis of rotation. We shall use this new reduced-mass, one-particle model of the diatomic rigid rotator to calculate allowed rotational energies and allowed transitions between those energies. We write equation

20 MOLECULES AND MOLECULAR LASERS FOR ELECTRICAL ENGINEERS

(2.30) for the model, defining the potential energy V(r), as zero for a torque-free rotator, so $H = p^2/2\mu$ and

$$HU(\underline{r},\phi) = E\ U\ (\underline{r},\phi)\ . \tag{3.7}$$

Recall that the momentum, p, is treated as an operator and H becomes

$$H = -\frac{h^2}{2\mu} \nabla^2 (\underline{r},\phi)\ .$$

In this model the only variable is ϕ, so ∇^2 is written in cylindrical coordinates, and equation (3.7) is

$$\frac{-h^2}{2I} \frac{\partial^2 U}{\partial \phi^2} = EU(\phi)$$

or

$$\frac{-h^2}{2I} \frac{\partial^2 U}{\partial \phi^2} = EU$$

$$\frac{\partial^2 U}{\partial \phi^2} = -\left[\frac{2IE}{h^2}\right] U \equiv (a)^2 U\ . \tag{3.8}$$

This last equation is satisfied by a function $U(\phi) = Ae^{\pm ja\phi}$, A a constant and $\psi = Ae^{\pm ja\phi} \cdot e^{-jEt/k}$.

The probability density is $\psi\psi^* = AA^*\ e^{ja\phi} \cdot e^{-ja\phi} = AA^* = A^2$, and the probability of finding the rotor within the angle $d\phi$ around the orientation ϕ is the product $\psi\psi^* d\phi$.

The wave function is now normalized, i.e.,

$$\int_a^b \psi\psi^* d\phi = 1 = A^2 \int_0^{2\pi} d\phi = 2\pi A^2$$

or

$$A = \frac{1}{\sqrt{2\pi}}\ . \tag{3.9}$$

Ordinarily, if A is real it does not alter the physical results obtained from operating on the wave function, so A suffices for both forms ($\pm j$) of the solution to equation (3.8); $U(\phi) = 1/\sqrt{2\pi}\ e^{\pm ja\phi}$.

The wave function must be single-valued, finite, and continuous. Either solution, $U(\phi)$, is continuous with a finite absolute value. The eigenfunction must be single-valued such that

$$U(\phi) = U(2\pi + \phi) = Ae^{ja\phi} = Ae^{ja(\phi+2\pi)} \tag{3.10}$$

$$e^{j2a\pi} = 1$$

where "a" is an integer. Therefore the two forms of the solution are

$$U = 1/\sqrt{2\pi}\ e^{ja\phi}, a = 0, \pm 1, \pm 2, \text{etc.}$$

The integers "a" are called the *quantum numbers* of the rotator and the energies going with the different values of "a" are

$$E = \frac{a^2 h^2}{2I}. \tag{3.11}$$

The *rotational term* F is defined as

$$F = \frac{E}{hc} \ (cm^{-1}) \tag{3.12}$$

so

$$F = \frac{a^2 h^2}{2Ihc} = \frac{a^2 h}{8\pi^2 Ic} = a^2 B \ (cm^{-1}) \tag{3.13}$$

where $B \equiv h/8\pi^2 Ic$, the rotational constant of the rotor. There are only *discrete values* of rotational energy, the *eigenvalues*—the F-values are not evenly spaced, i.e., 0, B, 4B, 9B, 16B, etc. The rotational term, F, is another way of writing energy and it is a more convenient unit to use in the laboratory. Note that the rotational constant B depends upon the moment of inertia of the molecule, so it is different for different types of molecules. Also, if we allow the interatomic spring in the model to stretch as the rotational energy increases, then B becomes a function of rotational energy.

Note that there are two eigenfunctions for each value of "a" except for $a = 0$, so the eigenfunctions are said to be "two-fold degenerate". In general, if there are n eigenfunctions which go with the same eigenvalue, then these eigenfunctions are "n-fold degenerate".

3 ABSORPTION AND EMISSION OF ENERGY BY THE RIGID ROTOR

The energy levels of the rigid rotor model have been determined, and now we must determine what frequencies of electromagnetic radiation will be absorbed or emitted by the diatomic molecule. These frequencies correspond to *changes* in energy of rotation, i.e., $\Delta E \sim F_1 - F_2$, and it will be necessary to learn what changes are allowed. The *selection rules* govern the transitions between the energy levels; these rules may be calculated in several ways.

We shall derive the selection rules from the Bohr Correspondence Principle (stated earlier) according to which the quantum and classical predictions should be the same for large values of the rotation quantum number, a.

If we study a rotating diatomic molecule or a rotating linear molecule using classical electromagnetic theory, the molecule must radiate at a frequency equal to the rotation frequency, provided that it has no electric dipole moment. The angular momentum L is

$$L = 2\pi \nu I = \omega I$$

and solving for ν, the frequency of rotation is

$$\nu = L/2\pi I, \tag{3.14}$$

a classical expression. In our model, L is really the z-component of angular momentum (rotation in x-y plane) and on the atomic scale the z-component of angular momentum is *quantized*, having values of ah. Therefore

$$v = \frac{L_z}{2\pi I} = \frac{ah}{2\pi I}. \tag{3.15}$$

The frequencies emitted or absorbed by the rotating diatomic molecule correspond to changes in its rotational energy, and according to the Bohr frequency relation the frequency associated with a transition between the a and a' rotation energy levels is

$$v = \frac{E_a - E'_a}{h} = \frac{(a^2 - a'^2)h}{8\pi^2 I}$$

$$= \frac{(a - a')(a + a')h}{8\pi^2 I}. \tag{3.16}$$

The Correspondence Principle says that the quantum number "a" must be large enough that the energy becomes of macroscopic order, i.e., the quantum situation approaches the classical situation, so set equation (3.15) equal to (3.16). As the two situations approach each other, it becomes increasingly difficult to distinguish between the levels a and a', and so $a + a' \cong 2a$. Thus

$$\frac{ah}{2\pi I} = \frac{2a(a - a')h}{8\pi^2 I} \tag{3.17}$$

and in order to preserve this equality, $(a - a')$ must be equal to one. Therefore the selection rule for rotation is

$$\Delta a = \pm 1, \tag{3.18}$$

that is, changes in rotational energy occur between adjacent levels for very large values of the quantum number, a. If we assume that the selection rule can be extrapolated backwards to the range of small values of a, then the selection rule for the rigid rotator is the same, and the frequencies of emission or absorption are predictable if the moment of inertia of the molecule is known. A more rigorous calculation of the selection rules for the planar rigid rotator yields the same results.

The first four rotation energy levels of the rotator are shown in Fig. 3.4, and the transitions indicated are those permitted by the selection rule $\Delta a = \pm 1$. The positive sign indicates absorption and the negative sign means emission. Note that the levels are unequally spaced and the energy differences ΔE_1, ΔE_2, and ΔE_3 are unequal. However, the differences $\Delta E_2 - \Delta E_1$ and $\Delta E_3 - \Delta E_2$ *are* equal. You will find these facts are true also for larger values of a.

Consider absorption of electromagnetic energy: the frequency of a rotation absorption is obtained by taking the difference (permitted by the selection rule) of two term values. We take the difference between the term value F' for a rotation state and the term value F'' for the next lower rotation state and we write

Frequency of spectral line $= \sigma = F' - F'' = [(a')^2 - (a'')^2]B$ (cm^{-1}) (3.19)

$\Delta a = +1$ for absorption

The first three spectral absorptions will occur at B, 3B, and 5B cm^{-1}. The spacing of the lines is constant, 2B cm^{-1}. As an example, a typical moment of inertia is $I \approx 10^{-39}$ $gm-cm^2$, and $2B = 5.58$ cm^{-1}. Therefore, depending upon the moment of inertia, one expects to see pure-rotation absorption (or emission) of electromagnetic energy by diatomic or linear molecules at frequencies in the microwave and/or the far-infrared region of the electromagnetic spectrum (see Table 1).

Any rotating molecule interacts with an electromagnetic field to produce an absorption or emission spectrum only if it has a nonzero electric dipole moment. The diatomic molecule H_2 has no rotation emission or absorption spectrum because it has a zero electric dipole moment, while HCl has a finite electric dipole moment and hence gives rise to rotation absorption and emission. Thus rotation spectra are associated with molecules having finite electric dipole moments; dipole moments exist because the center of charge of the molecule is not coincident with its center of mass. Such non-coincidence is the condition required for an n-pole electric moment (n = 2 for electric dipole).

4 A CALCULATION FROM ELEMENTARY PHYSICS

In elementary physics it is found that the average energy for a molecule per degree of freedom is $\frac{1}{2} kT$, where k is Boltzmann's constant ($k = 1.38 \times 10^{-16}$ $erg\text{-}deg^{-1}$) and T is the absolute temperature. The rotational kinetic energy is $\frac{1}{2} I \omega^2$. An average frequency of rotation, ν, can be found by equating these two expressions and solving for ν.

$$\nu = \frac{1}{2\pi} \sqrt{\frac{kT}{I}} \ rev/sec. \tag{3.20}$$

At room temperature $T = 300°K$, $\nu = 10^{12}/\pi$ rev/sec., or about 10 cm^{-1}, a wavelength of 1 mm (see Table 1).

5 THE USE OF PURE-ROTATION ABSORPTION AS A FILTER

A periodic pattern of energy absorption is often called a "comb" filter, and notice in Fig. 3.2 that the absorption spectrum of the rigid rotor is such a filter. Filtering at infrared and visible frequencies is desirable for many reasons, and it is one of the many jobs that atomic and molecular systems do extremely well and at relatively low cost. For each spectral line, it is possible to make reasonable predictions of absorption intensity, bandwidth, and bandshape from quantum theory; experimental checks of the calculations support the validity of the predictions.

The intensity of any spectral line is a superposition of intensities from transitions of the same kind in many molecules. The intensity of a single transition is proportional to the population *difference* which exists between the energy levels involved. More will be said about intensities later.

The change in the line shape of rotation absorption of a gas is due mainly to pressure changes. The effect is called pressure broadening because it is very nearly

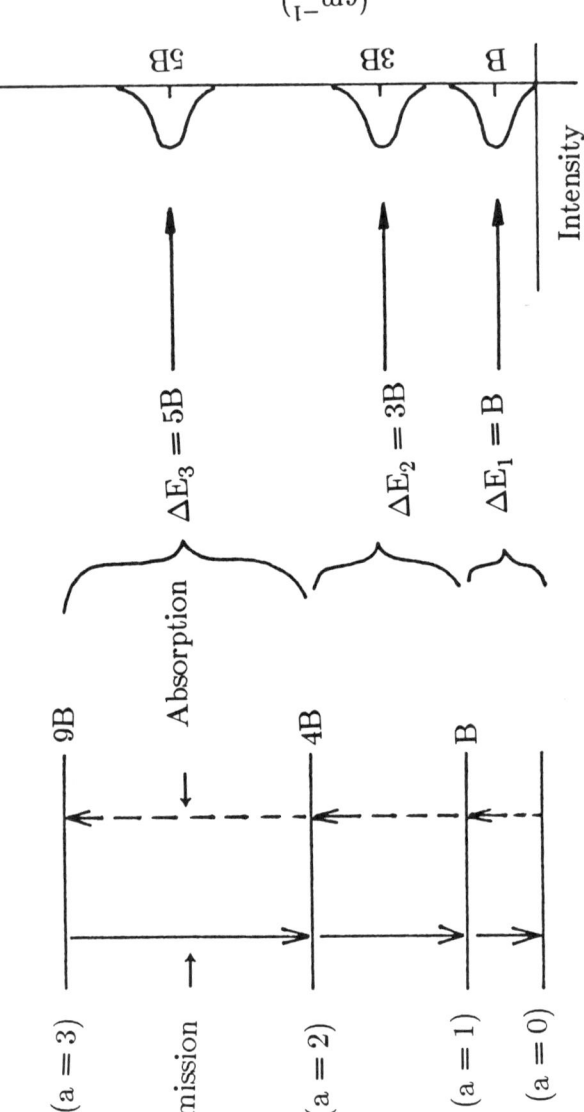

Figure 3.2 Pure-rotation energy levels of a rigid planar rotator.

proportional to pressure. At average temperature and pressure, gas molecules collide about 10^{10} times per second and between collisions they may rotate several thousand times. If the gas temperature is lowered, the collisions occur less frequently and the molecules rotate more slowly; hence there is practically no change with temperature in the number of rotations between collisions. To a very good approximation, a rotation absorption by a gas at low pressure has a characteristic Lorentz shape, just like the response of a series-tuned circuit at radio frequencies.

6 THE ROTATION ABSORPTION SPECTRUM OF A PRECESSING DIATOMIC MOLECULE

The rigid rotator has a fixed internuclear distance and a non-precessing axis. A real diatomic molecule, however, is free to precess in space: its axis of rotation may be oriented in any direction and so the theory of the rigid rotor must be modified. We consider now two angles:
(a) θ—angle describing rotation of the molecule about a given axis
(b) ϕ—angle describing the precession of the given axis about some space-fixed line.

These angles are shown in Fig. 3.3. From classical mechanics we learn that the angular momentum of the model is constant if there is no external torque. There is no external torque on the molecule if there is no external field present.

It is convenient to discuss the angular momentum, L, and its z-component, L_z. Solution of the Schrodinger time-independent equation using operator L yields eigenvalues of this operator which are

$$E = J(J+1)\,hcB\,, J = 0, 1, 2, 3, 4, \cdots \tag{3.21}$$

and the term value is now

$$F = E/hc = J(J+1)B \tag{3.22}$$

instead of $F = a^2 B$ for the rigid planar rotator. The frequency of a rotation absorption line is obtained by taking a difference of rotation terms. The selection rule is $\Delta J = \pm 1$, so we take the difference between the term value F' for an allowed rotation state and the term value F'' for the next lower allowed rotation state, i.e.,

$$\sigma = F' - F'' = J'(J'+1)B - J''(J''+1)B\,, J' = J'' + 1$$

$$\sigma = 2(J''+1)B \; (cm^{-1}). \tag{3.23}$$

Let $J'' = 0, 1, 2$, etc. and $\sigma = 2B, 4B, 6B$, etc., and the spacing between successive spectral lines remains 2B. The energy eigenvalues E_z, which go with L_z are found by choosing L_z as the operator in the Schrodinger time-independent equation.

$$E_z = Mh\,, M = 0, \pm 1, \pm 2, \pm 3, \cdots \pm J \tag{3.24}$$

Selection rule $\rightarrow \Delta M = 0, \pm 1$

The quantum number M does not occur unless there is an external force field acting upon the molecule; then M specifies the component of L along the space-fixed axis

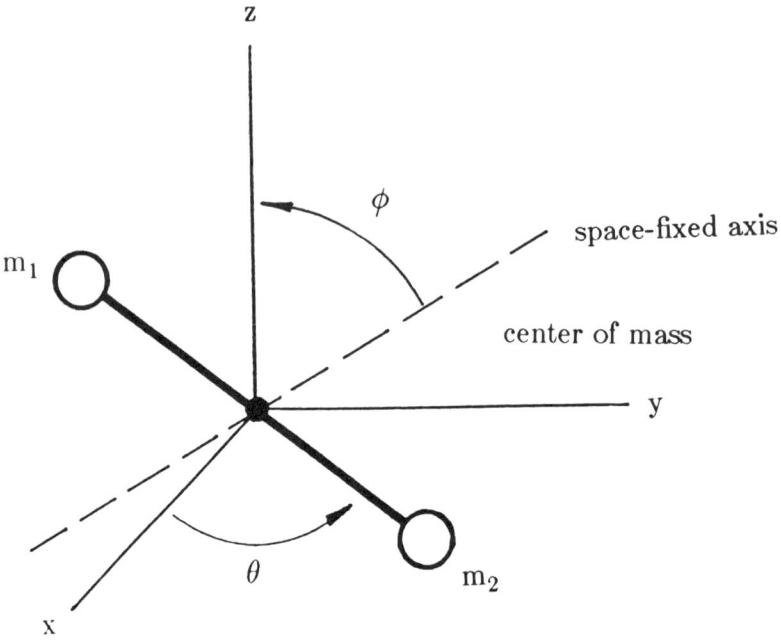

Figure 3.3 A rotating, precessing diatomic molecule.

defined by the direction of the external field. For each value of J, there are $2J + 1$ possible values of M, and the Jth eigenvalue is said to be $(2J + 1)$-fold degenerate. The relative population of the rotation energy levels is affected by this degeneracy and hence the intensity of an absorption (emission) depends upon J. The spectral position of the line remains constant, however, unless a strong external electric field exists.*

7 RELATIVE INTENSITIES OF ROTATION LINES

In classical physics, the Boltzmann energy distribution relates the number of molecules dN with energies between E and $E + dE$ to the absolute temperature, T, by

$$dN \approx e^{-E/kT}. \tag{3.25}$$

To relate this classical expression, which contains a continuous distribution of energy, to the quantum case, it is assumed that each energy state is proportional to $e^{-E/kT}$, ignoring energies that are not allowed by the Schrodinger equation. Recall that each rotation energy state described by J is $(2J + 1)$-fold degenerate, and so the relative number of molecules in the Jth energy level, N_J, is proportional to $(2J + 1)e^{-E/kT}$. The energy E is replaced by the rotation energy eigenvalue, so

---

*For more information on how rotation lines are moved and split by strong external electric fields see *Molecular Structure, The Physical Approach* by J. C. D. Brand and J. C. Speakman, Edward Arnold (Publishers) Ltd., London, 1960.

$$N_J \approx (2J + 1)e^{\frac{-J(J + 1)Bhc}{kT}} . \tag{3.26}$$

The total number of molecules in a sample is $N_t = \sum_J N_J$, so the population of the Jth energy state relative to the total population of a gas is

$$\frac{N_J}{N_t} = \frac{(2J + 1)e^{-J(J + 1)hcB/kT}}{\sum_J (2J + 1)e^{-J(J + 1)hcB/kT}} . \tag{3.27}$$

The sum goes from zero to infinity and the denominator can be evaluated approximately by an integral

$$N_t \approx \int_0^\infty (2J + 1)e^{-J(J + 1)hcB/kT} \, dJ = \int_0^\infty e^{-xhcB/kT} \, dx, \, x = J(J + 1)$$

$$= kT/Bhc . \tag{3.28}$$

Therefore

$$\frac{N_J}{N_t} \approx Bhc \, \frac{(2J + 1)e^{-J(J + 1)hcB/kT}}{kT} . \tag{3.29}$$

The relative intensities of rotation lines are

(a) for emission: $I_{em} = C' \, \dfrac{B}{T} \, \sigma^4 (2J' + 1)e^{-J'(J' + 1)hcB/kT}$ (3.30)

(b) for absorption: $I_{ab} = C'' \, \dfrac{B}{T} \, \sigma (2J'' + 1)e^{-J''(J'' + 1)hcB/kT}$ (3.31)

where C' and C'' are constants. The most intense absorption line in a rotation spectrum may be determined by differentiating I_{ab} with respect to J and equating the derivative to zero.

$$\frac{\partial I}{\partial J''} = \frac{\partial}{\partial J''} \left[C'' \, \frac{B^2}{T} \, (2J'' + 1)^2 \, e^{-J''(J'' + 1)hcB/kT} \right] = 0$$

Solving for J'',

$$J'' = \sqrt{\frac{kT}{hcB}} - \frac{1}{2} \tag{3.32}$$

and equation (3.32) gives the J-value of the lower rotation energy state from which the most intense line originates. Notice that this J-value depends upon both absolute temperature and the moment of inertia of the molecule. For light diatomic molecules such as HCl, the most intense rotation line occurs in the far infrared, but for heavier linear molecules the most intense line usually occurs in the microwave region.

8 CENTRIFUGAL STRETCHING OF THE INTERNUCLEAR DISTANCE

The internuclear distance in a real linear molecule is influenced by the centrifugal force due to rotation, so the internuclear distance increases with rotation energy. The effect of the stretching is to decrease slightly the frequency, σ, of the spectral line and the frequency of the free-axis rigid rotor is modified to allow for centrifugal stretching of the internuclear distances and for Coriolis forces. The rotational terms in a given vibrational level are calculated to be

$$F_v(J) = B_v J (J + 1) - D_v J^2 (J + 1)^2 + H_v J^3 (J + 1)^3 \qquad (3.33)$$

where D_v accounts for centrifugal stretching and H_v accounts for the Coriolis force which is due to the interaction of vibrational and rotational motions of the molecule. As a rotating diatomic molecule stretches, its rotation is slowed down and as the molecule contracts its rotation is speeded up by Coriolis forces. Normally the cubic term is not necessary in the analysis of an actual pure-rotation spectrum but it may happen, for example, in HF that this term must be included to explain the correct position of lines arising from transitions between the higher rotation energy levels. The absorption frequencies which are obtained from the selection rule $\Delta J = \pm 1$ are

$$\sigma_m = 2B_o (J + 1) - 2(2D_o - H_o)(J + 1)^3 + 6H_o (J + 1)^5 (cm^{-1}) \, . \qquad (3.34)$$

More complicated situations occur in the rotation of polyatomic molecules which are symmetric or asymmetric tops. Discussions of these cases are beyond the scope of this text, but the reader may find excellent discussions of these cases in Herzberg.*

9 EXAMPLE OF ENGINEERING USES OF PURE-ROTATION TRANSITIONS IN MOLECULES

An important example of the engineering use of pure-rotation transitions in molecules is laser emission from rotational transitions.

In 1967 Deutsch observed laser emission on pure-rotational transitions of hydrogen fluoride (HF) which was formed in a chemical reaction initiated by a pulsed electrical discharge. This was the first case of optical laser action on identified rotational transitions, and laser emission occurred at more than 24 frequencies in the spectral region 10-22 µm. Most of the observed transitions have J-values greater than 15 and they are dependent upon the type of gas used.

Laser action was also observed on nine HCl^{35} rotational transitions with wavelengths between 16.2 and 25.3 µ. The transitions identified in this region were in n = 0 (J = 19, 20, 30, 31, 32); n = 1 (J = 20, 21); and n = 2 (J = 20, 21). The laser emissions were excited in the same pulsed system using CH_3Cl, $CH_3Cl + Br_2$, and $CH_3Cl + Cl_2$ mixes.

---

*G. Herzberg, *Molecular Spectra and Molecular Structure*, D. Van Nostrand Co., Inc., Princeton, NJ, 1945.

Coleman[1] has written an excellent review paper on far-infrared molecular lasers and has done extensive work in developing practical laser sources in the far infrared. A paper by Hassler, Hubner and Coleman[2] is an example of lasing in the far infrared using a polyatomic molecule, SO_2. In particular this paper should drive home one *very important* idea to the student—one must know a great deal about the energy levels, selection rules, and relaxation rates of molecules (or atoms). This knowledge is requisite to success in designing a laser. Usually the spectroscopic groundwork is laid before an attempt is made at constructing a working laser.

PROBLEMS—PURE ROTATION

1. Show that the eigenfunctions $U_1 = Ae^{j\ell\phi}$ and $U_2 = Be^{-j\ell\phi}$ both satisfy the Schroedinger time-independent equation for the rigid diatomic rotator if A and B are constant and ℓ is an integer.
2. Let the instantaneous electric dipole moment of a rigid diatomic rotator be $M_y = M \cos\phi$, as seen by an observer. Show that the selection rule for this rotator is $\Delta J = \pm 1$.
3. Compute the electric dipole moment of $C^{12}O^{16}$. Assume an internuclear distance of 1.128 Å (1 Angstrom = 10^{-8} cm). The net charge on each atom is equal in magnitude to two electronic charges.
4. Repeat the calculation in problem three for HCl. Assume an internuclear distance of 0.958 Å. Let the net charge on each atom be one electronic charge (magnitude).
5. (a) Compute the moment of inertia of $C^{12}O^{16}$ using data from problem three. Show that $B = 1.93\ cm^{-1}$.
   (b) Calculate the numerical values of the rotational term F for $J = 0$ to $J = 4$.
   (c) Draw the energy level diagram in cm^{-1} and show beside it the pure-rotation absorption lines which correspond to the first four transitions.
6. Repeat problem 5 for HCl^{35}. Let $B = 10.59\ cm^{-1}$.
7. For HCl^{35}, write a computer program to evaluate the fraction $\dfrac{N_J}{N_t}$ of molecules in each rotational state from $J = 0$ to $J = 8$ at room temperature. Plot the fraction $\dfrac{N_J}{N_t}$ vs. J.
8. Derive an expression for J_{max}, the level from which the most intense pure rotation line should originate. Calculate this value for HCl^{35} and compare your answer with the plot of problem 7.

REFERENCES

1. Paul D. Coleman, "Far Infrared Molecular Lasers," IEEE J. of Quantum Electronics, *QE-9*, 130 (1973).
2. J. C. Hassler, G. Hubner, and P. D. Coleman, "Excitation Mechanism of the Far-Infrared Sulfur Dioxide Molecular Laser," J. Appl. Phys. *44*, 795 (1973).

CHAPTER

4

A VIBRATING DIATOMIC MOLECULE

1 INTRODUCTION

A diatomic molecule is modeled again as a mass-spring system to consider its vibration. The system has a natural resonance which depends upon the masses and the stiffness of the spring. The problem is approached from a classical point of view to illustrate the meaning of *anharmonicity* (or nonlinearity).

2 HOOKE'S LAW AND THE EFFECT OF ANHARMONICITY

In elementary physics one learns of Hooke's Law for springs. Such springs produce a restoring force F which is proportional to the displacement of the spring, $(x - x_e)$, or

$$F = -k_1(x - x_e), \qquad (4.1)$$

where k_1 is the spring constant, x_e is the equilibrium position of one end, and x is the position of that end when stretched. No real springs obey this law exactly and higher order terms are added to account for this discrepancy, i.e.,

$$F = -k_1(x - x_e) - k_2(x - x_e)^2 - k_3(x - x_3)^3 - \dots \qquad (4.2)$$

where $k_1 > k_2 > k_3 > ... k_n$. The force is assumed to be conservative, that is, it may be derived from a potential energy according to

$$F = -\nabla V. \tag{4.3}$$

Thus for a spring which obeys Hooke's Law, the force is

$$F = -k_1(x - x_e) = -\nabla V = -\frac{\partial V}{\partial x} \tag{4.4}$$

and

$$V = \frac{1}{2} k_1 (x - x_e)^2. \tag{4.5}$$

The total energy of this mass-spring system is

$$H = \frac{1}{2} \mu \dot{x}^2 + \frac{1}{2} k_1 (x - x_e)^2 \tag{4.6}$$

$$= \frac{P_x^2}{2\mu} + \frac{1}{2} k_1 (x - x_e)^2.$$

We write and solve the differential equation of motion for the spring described by Eq. (4.1). Using Newton's second law,

$$\mu \ddot{x} = -k_1 (x - x_e). \tag{4.7}$$

The solution is the familiar one for a *harmonic* oscillator

$$x - x_e = A \cos(\omega t + \theta), \tag{4.8}$$

where the constants A and θ depend upon the initial conditions and the frequency of oscillation is given by

$$f = \frac{1}{2\pi} \sqrt{\frac{k_1}{\mu}} \quad (cps) \tag{4.9}$$

or

$$\omega^2 = \frac{k_1}{\mu}. \tag{4.10}$$

Newton's second law is written again, but now there is included in the force the k_2 term from Eq. (4.2), i.e.,

$$\mu \ddot{x} = -k_1 (x - x_e) - k_2 (x - x_e)^2 \tag{4.11}$$

or

$$\ddot{x} = -\frac{k_1}{\mu}(x - x_e) - \frac{k_2}{\mu}(x - x_e)^2.$$

Substitute from Eq. (4.10) and

$$\ddot{x} = -\omega^2 \chi + \alpha \chi^2 \tag{4.12}$$

where $\chi = x - x_e$ and $\alpha = -\dfrac{k_2}{\mu}$. The solution is Eq. (4.12) is

$$x - x_e = A^2 \cos \omega t - \frac{\alpha A}{8\omega^2} \cos 2\omega t. \tag{4.13}$$

Comparison of Eqs. (4.8) and (4.13) shows the effect of anharmonicity in the restoring force of the spring. An *extra term* appears in Eq. (4.13): this extra term includes oscillation at twice the fundamental frequency but its amplitude is much smaller than the amplitude of the fundamental frequency. Thus for restoring forces which look like Eq. (4.2), we expect to see additional frequencies of oscillation besides the fundamental, and these additional frequencies are due to the nonlinearity (anharmonicity) of the restoring force.

Nonlinearities are familiar to electrical engineers: transistors and vacuum tubes may be operated as small-signal, linear devices, or as non-linear devices for use as harmonic generators, depending upon loading, biasing, etc. We shall see that molecules behave in much the same fashion.

If a vibrating molecule contains electric charges in such a configuration that a vibrating electric dipole moment occurs, then this vibrating dipole moment should radiate electromagnetic energy at the vibration frequency. Anharmonic vibrations should produce additional frequencies at smaller amplitudes. Such frequencies are in fact observed for vibrating molecules. If the electric charges are spatially oriented in a vibrating molecule in such a way that no net change occurs in the electric dipole moment, then the molecule has no emission or absorption spectrum in the infrared, but it is still possible that the molecule will *scatter* radiation in such a manner that another type of spectrum called the Raman spectrum is produced. This type of spectrum will be discussed in more detail later in Chapter 6.

3 QUANTUM MECHANICAL TREATMENT OF A HETERONUCLEAR DIATOMIC OSCILLATOR

The quantum mechanical treatment of a heteronuclear diatomic oscillator shows that not just any vibrational energy is allowed. The problem is solved as follows. The Schrodinger time-independent equation (2.30) is written using the total energy of Eq. (4.6) as the energy operator; anharmonicity is not considered now in the energy operator.

$$H \, U_n(x) = E_n \, U_n(x) \tag{2.30}$$

$$\left[\frac{P_x^2}{2\mu} + \frac{1}{2} k_1 (x - x_e)^2 \right] U_n(x) = E_n U_n(x) \tag{4.14}$$

The operator for P_x is $-\dfrac{h}{i} \dfrac{\partial}{\partial x}$, and so Eq. (4.14) becomes the differential equation

$$\frac{-h^2}{2\mu} \frac{\partial^2 U_n(x)}{\partial x^2} + \frac{1}{2} k_1 (x - x_e)^2 \, U_n(x) = E_n \, U_n(x) \, . \tag{4.15}$$

This equation is rewritten as

$$\frac{d^2 U_n(x)}{dx^2} + \frac{1}{2} k \left[\frac{-2\mu}{\hbar^2} \right] (x - x_e)^2 U_n(x) = \frac{-2\mu}{\hbar^2} E_n U_n(x) \qquad (4.16)$$

or

$$\frac{d^2 U_n}{dy^2} + (\lambda - y^2) U_n = 0 \qquad (4.17)$$

where

$$Y \equiv \sqrt[4]{k\mu/\hbar^2} \, x \, , \; \lambda \equiv \frac{2E}{\hbar\omega} \, , \; \omega \equiv \sqrt{\hbar/\mu} \, .$$

The solutions of this equation are of the form

$$U_n = \sum_0^N a_n \, y^n \, e^{-y^2/2} \qquad (4.18)$$

$$E_n = (n + 1/2) \hbar\omega \, , \; n = 0, 1, 2, \cdots \qquad (4.19)$$

The different allowed energy levels of the diatomic harmonic oscillator are generated by assuming different values for the vibrational quantum number n. Thus $E_1 = \frac{\hbar\omega}{2}$, $E_2 = \frac{3}{2} \hbar\omega$, $E_3 = \frac{5}{2} \hbar\omega$, etc. Notice that these energy levels are equally spaced. If we use the Correspondence Principle to obtain the vibrational selection rule, then the classical and quantum predictions should be the same for large values of n, or $n' - n'' = 1$ for the states of large energy. The extrapolation of this result to small values of n means that $\Delta n = \pm 1$, and vibration transitions are permitted between adjacent vibrational energy levels only.

We define a vibrational term in the same manner as we defined the rotational term, i.e.,

$$G = \frac{E_n}{hc} \, (cm^{-1}) \qquad (4.20)$$

where the E_n are the eigenvalues of *vibrational* energy. In order to get the frequency which corresponds to transitions between vibrational energy states, we apply the vibrational selection rule $\Delta n = \pm 1$ and take the difference of two terms. Thus

$$\sigma_v = G' - G'' = [(n' + 1/2) - (n'' + 1/2)] \frac{\omega}{2\pi c}$$

$$\sigma_v = G' - G'' = [(n' + 1/2) - (n'' + 1/2)] \frac{\omega}{2\pi c} \qquad (4.21)$$

$$= (n' - n'') \frac{\omega}{2\pi c} = \frac{(1)\omega}{2\pi c} = \frac{1}{\lambda} \, (cm^{-1})$$

and it is seen that all transitions will produce a spectral line at the same frequency, σ_v, because Δn can only be ± 1 (the + sign is used for absorption).

4 EFFECTS OF ANHARMONICITY ON REAL MOLECULES

Real molecules do not exhibit only one spectral line, however. Overtones (harmonics) occur at frequencies progressively *lower* than integral multiples of the fundamental vibrational frequency. This fact means that the vibrational force field of a molecule is actually anharmonic, so the quantum mechanical treatment must be modified for an *anharmonic* potential function like

$$V = \frac{1}{2} k_1 (x - x_e)^2 + \frac{k_2}{3} (x - x_e)^3 + \frac{k_3}{4} (x - x_e)^4. \tag{4.22}$$

This modification of the potential function leads to a perturbation calculation which has the simple harmonic oscillator as its starting point. The eigenvalues which result from such a calculation are

$$E_n = hcG = hc \,[G_o + (n + 1/2)\, \sigma' + (n + 1/2)^2 \, G_{11} + ...] \tag{4.23}$$

where G_o and G_{11} both depend upon the coefficients k_2 and k_3 in Eq. (4.22).

The selection rule for an anharmonic vibrator is $\Delta n = 0, \pm 1, \pm 2, ...$, and this rule permits overtone vibrations to occur. From an inspection of this selection rule it is seen that the overtone transitions occur between vibration energy levels that are not adjacent. The term G_{11} turns out to be a negative number, and so the equally-spaced energy levels of the simple harmonic oscillator are modified. They become progressively *closer together* as the quantum number n increases. The overtone frequencies will not occur at exact integral multiples of the fundamental frequency of vibration but rather they tend to converge with progressively decreasing spectral intensity.

A good example of such behavior is the HCl molecule. This molecule is important in chemical lasers. Its fundamental vibrational frequency and overtones are listed below in Table 4.1, along with the quantum numbers of the lower energy state involved in each transition.

Table 4.1 Fundamental and Overtone Vibrations of HCl

n'	σ (observed)	$\Delta \sigma$	$\Delta^2 \sigma$
0			
		2885.9	
1	2885.9		103.8
		2782.1	
2	5668.0		103.1
		2679.0	
3	8347.0		103.3
		2575.7	
4	10,922.7		

The frequency separation of successive spectral lines is labeled $\Delta \sigma$. These separations may be calculated by taking differences of successive *term differences*. For example, if we want to calculate the difference in frequency between the lines which originate from lower states $n' = 0$ and $n' = 1$ we first find the frequency of each line

$$\sigma(n' = 0) = [G(1) - G(0)] \qquad (4.24)$$

$$\sigma(n' = 1) = [G(2) - G(0)] \qquad (4.25)$$

and take the difference

$$\Delta\sigma = \sigma(n' = 1) - \sigma(n' = 0) = [G(2) - G(1)] \; (cm^{-1}). \qquad (4.26)$$

(Table 4.1 tells us that $\Delta\sigma$ for this case is 2885.9 cm^{-1}.) This process may be repeated for the next higher state, $n' = 2$, etc. The second differences are labeled $\Delta^2\sigma$. The purpose in making such a table is to show the reader the effects of anharmonicity on the overtone frequencies and also to demonstrate how the constant G_{11} may be determined from such data. Eq. (4.26) may be written as

$$\Delta\sigma = [G(n+1) - G(n)] \qquad (4.27)$$

and if the expression for G is substituted we get

$$\Delta\sigma = [G_o + \sigma'(n + \frac{3}{2}) + (n + \frac{3}{2})^2 G_{11}] - [G_o + \sigma'(n + \frac{1}{2}) + (n + \frac{1}{2})^2 G_{11}]$$

$$= \sigma' + 2(n+1)G_{11} = (\sigma' + 2G_{11}) = 2nG_{11}. \qquad (4.28)$$

This is the equation of a straight line, and the value of G_{11} may be determined from experimental measurements of $\Delta\sigma$. Equation (4.28) is called a "combination" relation and such relations are used extensively to determine molecular parameters from experimental data. The value of G_{11} is determined from a plot like that shown in Fig. 4.1 The slope of the line is $2G_{11}$.

The second differences are calculated in a manner similar to Eq. (4.26) and the result is that $\Delta^2\sigma = 2G_{11} = -51.7 \; cm^{-1}$ for HCl. It follows that σ' may be found by using G_{11} in the expressions for the vibrational terms.

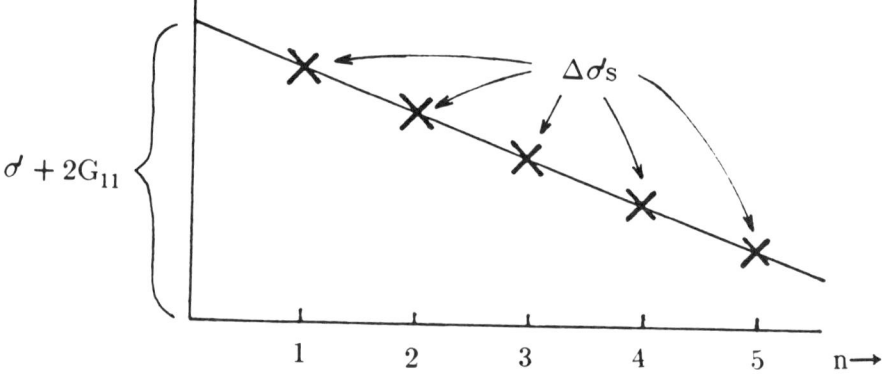

Figure 4.1 A typical plot of Eq. (4.28).

$$G(0 + 1/2) = 2885.9 = \sigma' + 2G_{11} \Longrightarrow \sigma' = 2989.3$$

$$G(1 + 1/2) = 2782.1 = \sigma' + 4G_{11} \Longrightarrow \sigma' = 2988.9$$

$$G(2 + 1/2) = 2697.1 = \sigma' + 6G_{11} \Longrightarrow \sigma' = 2989.3$$

$$G(3 + 1/2) = 2575.7 = \sigma' + 8G_{11} \Longrightarrow \sigma' = 2989.3$$

The average value of σ' is therefore 2989.2 cm^{-1}, and this value is more than 100 cm^{-1} higher than the observed fundamental frequency listed in Table 4.1. Such large deviations from harmonic oscillator values are common in molecules which contain a light atom such as hydrogen.

If one considers a polyatomic molecule instead of a diatomic molecule, certain linear combinations of the 3N-6 or 3N-5 fundamental frequencies may occur. These combinations may be sums or differences of fundamentals, thus adding to the complexity of the observed spectrum. (Later it will be shown how to determine selection rules for the overtones and combinations for a given molecular structure). Two other important types of vibration absorption bands may appear in the spectra of molecules; these absorptions arise from *Fermi resonance* and from *"hot"* bands. Let us examine each in turn.

5 FERMI RESONANCE

In polyatomic molecules it sometimes happens that two vibrational energy levels which belong to different vibrations have almost equal energies; such levels are said to be *accidentally degenerate*. Thus there is approximate "resonance" of these levels. Such a resonance leads to a disturbance or perturbation of the two levels and the energy spacing of the levels changes. Each level then is a mixture of the two, and experimentally this mixing is observed as *two* strong infrared absorptions rather than one. Also, it does happen that a sum or a difference of fundamentals (called a "combination" band) or an overtone band of a fundamental occurs near some other fundamental band. In these cases the overtone or combination band robs energy from the fundamental and enhances its own absorption intensity out of proportion to that which would normally occur otherwise.

6 HOT BANDS

Hot bands may best be understood by referring to Fig. 4.2. Two fundamental frequencies, v_1 and v_2, may combine to produce a combination frequency such as $(v_1 + v_2)$ or $(v_1 - v_2)$. A hot band is the result of a transition in which the energy difference corresponds to $(v_1 + v_2) - v_2$. The term "hot" is used because upon cooling a sample, this type of band will decrease in absorption intensity or disappear altogether. Cooling decreases the number of molecules in the energy state labeled 1, hence not as many molecules are available for that transition and the intensity of a hot band decreases rapidly with falling temperature.

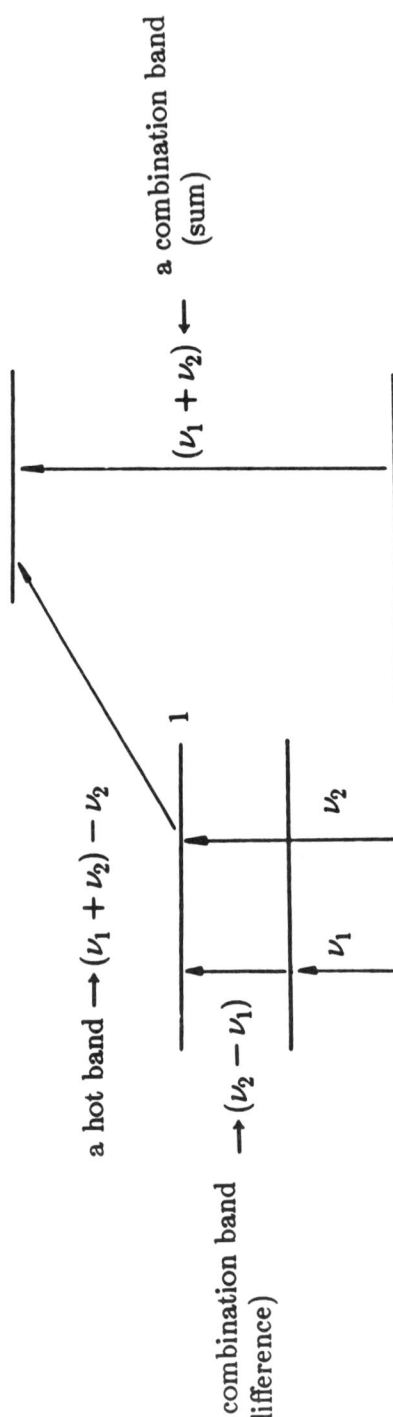

Figure 4.2 The origin of a hot band and of a combination band.

7 COMPARISON OF MOLECULAR RESONANCES AND RESONANCES OF DISCRETE CIRCUITS

At this point we pause in our discussions of the origins of molecular spectra to point out an important comparison between molecular resonances and resonances of discrete electrical circuits. The response of a molecule to an electromagnetic wave may be compared to the response of a series or parallel RLC circuit. Let us consider first the familiar frequency dependence of the impedance of a series RLC circuit and the frequency dependence of the admittance of a parallel RLC circuit.

For a series RLC circuit the impedance is

$$Z = R + j\omega L - \frac{j}{\omega C} = R + j\omega L \left[1 - \frac{1}{\omega^2 LC}\right] = R + j\omega L \left[1 - \frac{\omega_o^2}{\omega^2}\right]$$

$$= R + \frac{jL}{\omega}(\omega^2 - \omega_o^2) \tag{4.29}$$

where $\omega_o^2 = 1/LC$, the series resonant frequency. If we consider frequencies ω which are close to the resonant frequency then

$$\omega^2 - \omega_o^2 = (\omega + \omega_o)(\omega - \omega_o) \approx 2\omega(\omega - \omega_o) \tag{4.30}$$

and

$$Z = R + j2L(\omega - \omega_o) = R[1 + 2j\frac{L}{R}(\omega - \omega_o)]$$

$$= R\left[1 + 2j\frac{(\omega - \omega_o)}{\gamma_s}\right] \tag{4.31}$$

where $\gamma_s = \frac{R}{L} = \frac{1}{\text{time constant}} = $ damping rate (sec^{-1}). The current in the series circuit is

$$I = \frac{V}{Z} = \frac{V}{R} \cdot \frac{1}{1 + 2j\frac{(\omega - \omega_o)}{\gamma_s}}. \tag{4.32}$$

For the parallel case the following changes are made in the symbols in Eq. (4.31)

$R \to G$

$L \to C$

$C \to L$

$Z \to Y$.

This equation may be rewritten as

$$Y = G[1 + 2j\,RC\,(\omega - \omega_o)] = G\left[1 + 2j\frac{(\omega - \omega_o)}{\gamma_p}\right] \tag{4.33}$$

where now $\gamma_p = \dfrac{1}{RC} = \dfrac{1}{\text{time constant}} = $ damping rate (sec^{-1}). The frequency dependence of the voltage across the parallel RLC circuit is

$$V = \frac{I}{Y} = \frac{I}{G} \cdot \frac{1}{1 + 2j\dfrac{(\omega - \omega_o)}{\gamma_p}}. \tag{4.34}$$

The second factor in Eq. (4.34) or Eq. (4.32) may be written as $1/(1+j\delta)$ and the power response is proportional to $1/(1+\delta^2)$. This form is called the Lorentz lineshape and the previous form is called the complex Lorentz lineshape. The half bandwidth of the response curve is determined by setting

$$\left| \frac{1}{1+j\delta} \right| = \frac{1}{\sqrt{2}}. \tag{4.35}$$

Therefore $\delta = 1 = \dfrac{2(\omega - \omega_o)}{\gamma_p}$, or $\omega - \omega_o = \dfrac{\gamma_p}{2}$, dropping the subscript s or p. The bandwidth is therefore

$$2(\omega - \omega_o) = \gamma_p. \tag{4.36}$$

If we rewrite equation Eq. (4.32) terms of the quality factor, Q, then the response of the electrical circuit for frequencies close to ω_o is proportional to

$$\frac{1}{1 + 2jQ\dfrac{(\omega - \omega_o)}{\omega_o}} \text{ where } Q = \frac{\omega_o L}{R}.$$

Now let us consider a molecule as a resonant circuit. We have seen earlier that if a mode of vibration produces a net change in the electric dipole moment of the molecule the molecule will absorb infrared energy. For simplicity we consider the harmonic oscillator model of a heteronuclear diatomic molecule with a reduced mass μ. Let the molecule interact with a one-dimensional, time-varying electric field $E_x(t)$. If the molecular motion is damped by collision with neighboring molecules the classical differential equation of motion becomes

$$\mu \ddot{x} + D \dot{x} + kx = -eE_x(t) \tag{4.37}$$

where D is the damping due to both radiation of energy by the molecule and collisions with neighbors. Divide Eq. (4.37) by μ

$$\ddot{x} + \frac{D}{\mu} \dot{x} + \frac{k}{\mu} x = \frac{-eE_x}{\mu}(t) \tag{4.38}$$

and define $\dfrac{D}{\mu} \equiv$, $\dfrac{k}{\mu} \equiv \omega_o^2$. It is important to note that in the diatomic case, the actual absorption frequency for the fundamental mode of vibration ($\Delta J = 0$) corresponds very closely to $\dfrac{1}{2\pi \sqrt{k/\mu}}$.

Let $E_x(t)$ be represented by a Fourier integral

$$E_x(t) = \frac{1}{2\pi} \int_{-\infty}^{\infty} E_x(\omega) e^{j\omega t} d\omega. \qquad (4.39)$$

If this integral expression for $E_x(t)$ is substituted on the right side of Eq. (4.38) and if we assume that the solution of the differential equation is of the form

$$x(t) = \frac{1}{2\pi} \int_{-\infty}^{\infty} a(\omega) e^{j\omega t} d\omega \qquad (4.40)$$

then when we substitute this expression on the left side of Eq. (4.38) and perform the indicated differentiation with respect to time, the result is

$$\frac{1}{2\pi} \int_{-\infty}^{\infty} (-\omega^2 + j\ell\omega + \omega_o^2) a(\omega) e^{j\omega t} d\omega = \frac{-e}{\mu\sqrt{2\pi}} \int_{-\infty}^{\infty} E_x(\omega) e^{j\omega t} d\omega. \qquad (4.41)$$

Collect all terms on the left side and

$$\int_{-\infty}^{\infty} \left[\left[-\omega^2 + j\ell\omega + \omega_o^2 \right] a(\omega) + \frac{eE_x(\omega)}{\mu} \right] e^{j\omega t} d\omega = 0. \qquad (4.42)$$

This integral must be identically zero for all frequencies ω; therefore the integrand within the brackets must be identically zero and

$$G(\omega) \equiv (-\omega^2 + j\ell\omega + \omega_o^2) a(\omega) + \frac{eE_x(\omega)}{\mu} = 0.$$

Solve for $a(\omega)$:

$$a(\omega) = \frac{-eE_x(\omega)}{\mu} \cdot \frac{\ell}{(-\omega^2 + j\ell\omega + \omega_o^2)} = \frac{jeE_x(\omega)}{\mu} \cdot \frac{1}{\ell\omega + j(\omega^2 - \omega_o^2)}. \qquad (4.43)$$

For frequencies close to ω_o we make the same approximation as before and

$$a(\omega) \approx \frac{jeE_x(\omega)}{\mu\ell\omega} \cdot \frac{1}{1 + 2j\dfrac{(\omega - \omega_o)}{\ell}}. \qquad (4.44)$$

Notice that now the damping rate is designated by ℓ and if we compare Eq. (4.45) to Eqs. (4.32) or (4.34), then the molecular "bandwidth" is ℓ. Therefore

$$Q = \frac{\omega_o}{\ell} = \frac{\omega_o}{\Delta\omega}. \qquad (4.45)$$

There are many theories to explain the actual shape and width of spectral lines of molecules—lines which originate from rotation or vibration transitions. In laser applications it is important to know such information. Breene[1] discusses such theories in detail, and he presents some examples of attempts to verify experimentally the predictions of the theories.

---

[1] The Shift and Shape of Spectral Lines, R. G. Breene, Jr., Pergamon Press, Inc., N.Y., 1961.

In the following chapter is presented the origin of spectral lines which result from transitions between energy states related to *vibration-rotation*. Such spectral lines are very important in applications such as the CO_2 laser which is presented in Chapter 10 as an example of lasing on *vibration-rotation* transitions.

CHAPTER

5

VIBRATION—ROTATION

1 INTRODUCTION

The vibration and rotation of a diatomic molecule have been discussed as separate topics. Pure-rotation transitions in gases occur in the far infrared and microwave but "pure" vibration does not occur in a real molecule in the gas phase because a molecule rotates as it vibrates. The infrared selection rules allow vibration and vibration-rotation transitions; such transitions occur at higher energies than pure-rotation transition, *viz.,* in the medium infrared, the near infrared, and the photographic infrared. As a starting point for understanding these transitions, we will consider that the energies of vibration and rotation are independent but additive. This assumption is a zero-order approximation. It does not result in a complete interpretation of all of the features of a vibration-rotation transition because the moment of inertia of a molecule depends upon the value of the internuclear distance which changes with time during vibration. Also, centrifugal stretching of the internuclear distance affects the anharmonicity of the force field. The zero-order approximation is useful for gaining understanding, through, so let us examine it now.

2 ZERO-ORDER APPROXIMATION

The total energy of a vibrating and rotating molecule is assumed to be additive and it is written

$$E_{V+R} = hc(G + F) = hcT, \qquad (5.1)$$

where G is given by Eq. (4.20) with $n = 0,1,2,...$, and F is given by Eq. (3.13) with $J = 0,1,2,....$ The selection rules are $\Delta n = 0,\pm 1,+12,...$, and $\Delta J = \pm 1$. On the basis of the zero-order approximation, the energy levels through $n = 3$ are shown in Fig. 5.1. Transitions are permitted between vibration-rotation states and the frequency of resulting spectral lines is

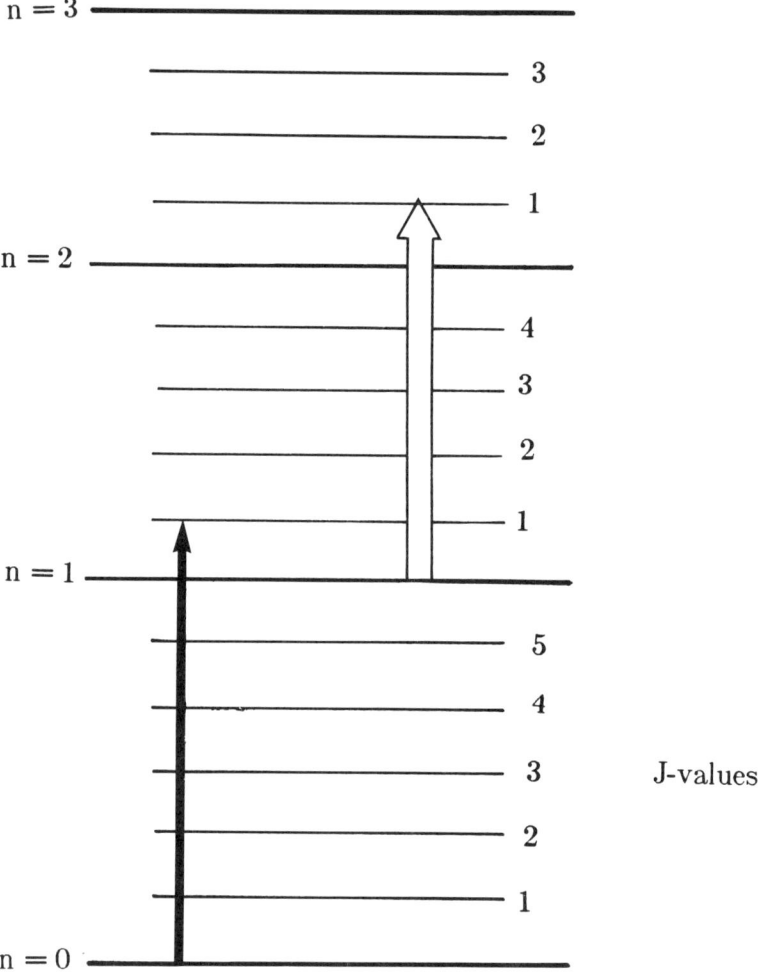

Figure 5.1 Energy levels for a vibrating, rotating diatomic molecule, zero-order approximation.

$$\sigma_{V+R} = T' - T''$$
$$= [G(n') + F(J')] - [G(n'') + F(J'')] \quad (5.2)$$
$$= G(n') - G(n'') + F(J') - F(J'')$$

and $\Delta n = n' - n''$, $\Delta J = J' - J''$. For example, let $\Delta n = 1$, $\Delta J = +1$; a transition which satisfies this rule is shown in the figure by a solid arrow. An equally valid choice is shown by the double arrow, but the intensity of this transition will be considerably weaker because it arises from a higher energy state with much less population than $n = 0$. At room temperature, about 99% of all molecules are in the $n = 0$ state.

3 THE VIBRATION-ROTATION BAND

A vibration-rotation band which is termed the "fundamental" mode of the molecule is that which arises from a fixed value of Δn, associated with *two* possible values of J. There are two "branches" or series of spectral lines which arise from the two possible values of J. These are

P-branch: $\Delta J = -1$, thus $J' = J'' - 1$

R-branch: $\Delta J = +1$, thus $J' = J'' + 1$.

The *band center* corresponds to $\Delta J = 0$ and therefore

$$\sigma_c = G(n') - G(n'') \quad (5.3)$$

and

$$\sigma_{V+R} = \sigma_c + F(J') - F(J'') . \quad (5.4)$$

The frequencies of the spectral lines in the P and R branches are

$$\sigma_p = \sigma_c + F(J'' - 1) - F(J'')$$
$$= \sigma_c + (J'' - 1) J'' B - J'' (J'' + 1) B$$
$$= \sigma_c - 2J'' B, \quad J'' = 1,2,3...$$
$$= \sigma_c - \Delta\sigma_1 \quad (5.5)$$
$$\sigma_R = \sigma_c + F(J'' + 1) - F(J'')$$
$$= \sigma_c + (J'' + 1)(J'' + 2) B - J''(J'' + 1) B$$
$$= \sigma_c + 2(J'' + 1) B, \quad J'' = 0,1,2,3,4,...$$
$$= \sigma_c + \Delta\sigma_2 . \quad (5.6)$$

The spectral band of frequencies which arise from these vibration-rotation transitions is shown in Fig. 5.2. Exactly how many lines are observed for a given type of molecule depends upon the intensity which depends, in turn, upon

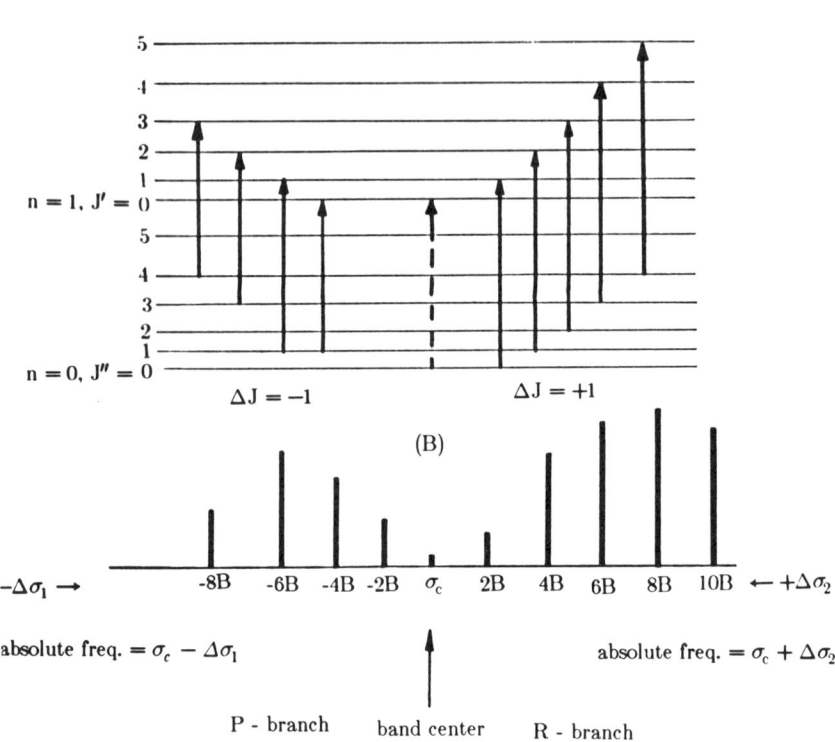

Figure 5.2 Origin of P and R branches in a vibration-rotation band, zero-order approximation. (A) Energy diagram showing vibration and rotation levels. (B) Spectral lines which result from each vibration-rotation transition shown in (A); intensities are approximate.

(a) the number of molecules in the initial state
(b) the energy carried off by a photon (hcσ)
(c) the probability of a given transition occurring.

If the vibration band is due to absorption of energy the line intensity depends upon the first power of the line frequency: for emission the intensity depends upon the fourth power of the frequency. For a diatomic molecule the transition probability is independent of J.

It was shown previously in Chapter 3 that the intensity of a rotational line is

$$N_j \approx (2J + 1) e^{-E/kT}.$$

If E becomes hcT, then the total number of molecules in a vibrational level is N_v, and the fraction of molecules in the Jth rotational state is

$$= N_v \cdot \frac{N_J}{N_t}.$$

$$N_{VJ} = N_v \frac{(hcB)}{kT} (2J + 1) e^{-J(J+1)Bhc/kT} \tag{5.7}$$

4 HIGHER-ORDER APPROXIMATIONS

The zero-order approximation is useful to explain the origin of the P and R branches in a fundamental vibration-rotation band. A first-order approximation which takes into account the variable internuclear distance modifies the equal spacing of the lines in these branches for the following reasons. During vibration the average internuclear distance is greater than the equilibrium distance, so the average moment of inertia of the molecule is greater than the moment of inertia at equilibrium. As a result, the rotational constant is smaller than the equilibrium value, and the effect of vibration upon B is greater for larger values of the vibrational quantum number. Therefore the rotational constant B becomes a function of n according to

$$B_n = B - \alpha(n + 1/2), \tag{5.8}$$

where B is the equilibrium value of the rotational constant and α is a function of the anharmonic constants.

To account for the centrifugal distortion of the molecule during rotation, the rotational term $F(J)$ is modified as

$$F(J) = J(J+1)B_n - J^2(J+1)^2 D_n \tag{5.9}$$

where D_n is given by

$$D_n = \frac{4B^3}{\sigma_o^2} - B(n + 1/2). \tag{5.10}$$

The reader will recall that rotational energy levels are not equally spaced, even when vibration is ignored. When the effects of vibration are considered, the rotational levels are spaced differently in different vibrational states, and the vibration-rotation lines are less widely spaced for larger values of the vibration quantum number, i.e., for overtones of the fundamental.

For the purpose of illustration, it will suffice to consider only $F(J) = J(J+1)B_n$ for the rotational term and calculate again the frequencies of the spectral lines in the P and R branches using Eq. (5.4).

$$\sigma_{V+R} = \sigma_c + F(J') - F(J'') \tag{5.4}$$

$$\sigma_p = \sigma_c + J'(J'+1)B_{n'} - J''(J''+1)B_{n''} \tag{5.11}$$

Notice that the rotation quantum number, J, *must* agree with the vibration *levels* (denoted by one or two primes) because now B_n is considered to be a function of n. Thus

$$\sigma_p = \sigma_c + J''(J''-1)B_{n'} - J''(J''+1)B_{n''}$$
$$= \sigma_c - (B_{n'} + B_{n''})J'' + (B_{n'} - B_{n''})J''^2, J'' = 1,2,3,...$$

and from Eq. (5.6) we can write

$$\sigma_R = \sigma_c + (J''+1)(J''+2)B_{n'} - J''(J''+1)B_{n''} \qquad (5.12)$$

$$= \sigma_c + (B_{n'} + B_{n''})(J''+1) + (B_{n'} - B_{n''})(J''+1)^2, J'' = 0,1,2,....$$

$$\sigma = \sigma_c - 2B_e J'' \text{ and } \sigma = \sigma_c + 2B_e(J''+1).$$

Comparison of Eqs. (5.11) and (5.12) with Eqs. (5.5) and (5.6) shows that the equilibrium rotational constant B has been replaced by $\frac{1}{2}(B_{n'} + B_{n''})$ and another term involving the difference $(B_{n'} - B_{n''})$. The effect of this difference on the spectral lines is very small and may only be seen for lines far from σ_c. The expressions for σ_P and σ_R are usually combined into one expression as follows,

$$\sigma = \sigma_c + (B_{n'} + B_{n''})m + (B_{n'} - B_{n''})m^2, \qquad (5.13)$$

where
$$m = -J'' \text{ for the P-branch}$$
$$m = J'' + 1 \text{ for the R-branch.}$$

The third term in Eq. (5.13) causes the lines in the R-branch to converge while the lines in the P-branch diverge. If the third term is large enough, the converging R-branch turns back on itself before large enough quantum numbers can occur to reduce the band intensity to near zero. The turning point is called a *band head*. Such behavior is depicted in Fig. 5.3 by a *Fortrat* diagram which is a plot of Eq. (5.13).

Line intensities are plotted below the diagram to illustrate further the convergence and turning back of the R-branch lines. Combination relations will be used again to distinguish the calculations of the constants B_n and D_n for the upper and lower vibrational

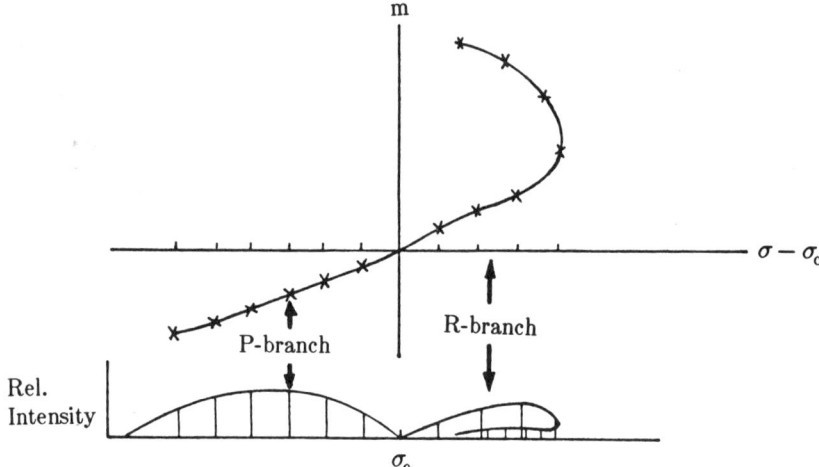

Figure 5.3 A Fortrat diagram showing convergence and band head of a typical fundamental vibration-rotation band of a diatomic molecule.

states. It is easier to use the combination relations because we want expressions for $\sigma_R - \sigma_P$ which involve spectral lines with a *common* rotational level.

$$\sigma_P = \sigma_c + F'(J-1) - F''(J) \left\{ = \sigma_c + F'(J) - F''(J+1) \right\} \tag{5.14}$$

$$\sigma_R = \sigma_c + F'(J+1) - F''(J) \left\{ = \sigma_c + F'(J) - F''(J-1) \right\} \tag{5.15}$$

The expressions in the brackets are equivalent expressions because ΔJ remains -1 for the P-branch and +1 for the R-branch. Subtract the bracketed expressions above.

$$\sigma_R - \sigma_P = F''(J+1) - F''(J-1) \equiv \Delta F''(J) \tag{5.16}$$

Now subtract the unbracketed expressions

$$\sigma_R - \sigma_P = F'(J+1) - F'(J-1) \equiv \Delta F'(J) . \tag{5.17}$$

Notice that Eqs. (5.16) and (5.17) are identical in the variable J but describe different rotational levels. The expression for $\Delta F(J)$ is

$$\Delta F(J) = \left[(J+1)(J+2)B_n - (J+1)^2(J+2)^2 D_n\right]$$
$$\quad - \left[J(J-1)B_n - J^2(J-1)^2 D_n\right]$$
$$= (J+1/2)(4B_n - 6D_n) - 8(J+1/2)^3 D_n \tag{5.18}$$

and so

$$\Delta F''(J) = (J+1/2)(4B_{n''} - 6D_{n''}) - 8(J+1/2)^3 D_{n''} \tag{5.19}$$

$$\Delta F'(J) = (J+1/2)(4B_{n'} - 6D_{n'}) - 8(J+1/2)^3 D_{n'} . \tag{5.20}$$

Eq. (5.18) may be written as

$$\frac{\Delta F(J)}{(J+1/2)} = (4B_n - 6D_n) - 8(J+1/2)^2 D_n \tag{5.21}$$

and then $\dfrac{\Delta F(J)}{(J+1/2)}$ is plotted against $(J+1/2)^2$. The intercept of the resulting straight line occurs at the average value $(4B_n - 6D_n)$ and the slope of the line is $-8D_n$.

4.1 Calculation of Q

We choose a vibration-rotation line in the fundamental band of HCl a line at 2325 cm^{-1}. Its measured bandwidth is about 2 cm^{-1}. Using Eq. (4.45) the Q is 1170 and so the energy stored per cycle is large.

PROBLEMS VIBRATION-ROTATION

1. Assume the wavelengths of the fundamental vibration and the first two overtone of $C^{12}O^{16}$ are $\lambda_1 = 4.66 \ \mu m$, $\lambda_2 = 2.35 \ \mu m$, and $\lambda_3 = 1.57 \ \mu m$, respectively.
   (a) Calculate the frequency of each transition in cm^{-1}.
   (b) Calculate the values of σ_o and G_{11}.
   (c) Calculate the force constant.

2. (a) Compute the values of the vibrational terms of HCl^{35} for $n = 0$ and $n = 1$. Assume $\sigma_o = 2988.95 \ cm^{-1}$, $2G_{11} = -51.65 \ cm^{-1}$; the rotational constants are $B_o = 10.44 \ cm^{-1}$ and $B_1 = 10.14 \ cm^{-1}$.
   (b) Calculate the energy (in cm^{-1}) of the first six rotational levels in the vibrational states $n = 0$ and $n = 1$.
   (c) Calculate the frequency of the center of the vibration-rotation band and also the frequencies of the first six lines which occur in the P and R branches.

3. Show that the most intense line in the P and R branches of a vibration-rotation band originates from the rotational level $J \ max = \sqrt{\dfrac{kT}{hcB}} - \dfrac{1}{2}$.

4. Go to the literature and find a paper by Whitcomb and Lagemann, *Physical Review*, Vol. 55, page 181, 1939, which contains results of an investigation of the rotational structure of the fundamental and first overtone of CO. Use the combination relations on their data and calculate the values of the rotational constants B_n and D_n for the vibrational states $n = 0$ and $n = 1$.

CHAPTER

6

THE RAMAN EFFECT

1 INTRODUCTION

The Raman effect is the inelastic scattering by molecules of photons of light. If a photon is scattered inelastically by a molecule it may gain energy and be scattered at a frequency higher than the original frequency, or it may lose energy and be scattered at a frequency lower than the original frequency. Thus weak spectral bands appear above and below the original frequency; these bands are called anti-Stokes or Stokes bands, respectively. These Raman spectral bands have very low intensity, about 10^{-11} of the incident light intensity, but such low intensities are observable with photo-electric detectors. The effect is independent of the frequency of the incident photon.

The selection rules for Raman scattering are different for each type of molecular symmetry, and these rules determine which Raman spectral bands are allowed. Also, the selection rules for Raman scattering differ from the selection rules for infrared absorption: this difference is used to advantage in studies of molecular structure, fundamental modes of vibration, or other physical and electrical properties of molecules because both infrared and Raman data may be brought to bear upon the same problem. The Raman spectral bands which result from monochomatic photon excitation of liquids arise from fundamental modes of molecular vibrations in the liquid, so the task of determining fundamental modes is simplified and aided by Raman data. The vibrational symmetry of these modes may often be inferred by measuring the polarization of the Raman light; the bands which are completely polarized (or nearly so) arise from modes of high symmetry while those which are only partially polarized arise from asymmetric modes (or nearly so).

Raman data are useful for the design of quantum electronic systems in which a laser is to be converted from one frequency to another, or in which a conversion as well as incremental tuning of the converted frequency is accomplished. Spatial, electrical, and optical properties of crystals and semiconductors are often studied using the Raman effect. The elucidation of crystal properties is aided by polarizing the incident monochromatic radiation with respect to the various crystal axes. Gas lasers, pulsed or cw, which operate in the visible region of the spectrum are commonly used to supply the incident radiation for Raman spectroscopy. High laser power is not required to observe the Raman effect. In most clear and slightly-colored liquids, power levels of 1-5 (mW) may be used. To observe the Raman effect in gases and solids, higher power ranging from 50 (mW)—5W is desirable (but not always necessary).

2 CLASSICAL THEORY OF THE RAMAN EFFECT

The classical theory of the Raman effect explains the frequency shifts which are observed experimentally. Consider monochromatic radiation which interacts with a vibrating molecule. As the molecule vibrates, its dipole moment $\bar{p}$, may change or its polarizability, α, may change or both. The dipole moment per unit volume is the polarization $\bar{P}$ and it is related to the polarizability by the electric field, $\bar{E}$, of the incident radiation field by

$$\bar{P} = \alpha \bar{E} . \qquad (6.1)$$

The polarizability may be a constant or a tensor, depending upon the properties of the medium. For the moment, we consider it to be a function of the internuclear distance, $\alpha(x)$. Suppose the molecule vibrates only along the x-axis and the vibrational amplitude is very small. For such a case the polarizability may be written as

$$\alpha(x) \approx \alpha_o + \frac{\beta x}{A} , \qquad (6.2)$$

a first order power series expansion of $\alpha(x)$. The constant α_o is the static polarizability of the molecule at equilibrium, β is the time rate of variation of the polarizability, and A is the vibration amplitude of the molecule. If the incident electric field is sinusoidal at the frequency ν_i, then

$$E = E_o \sin 2\pi \nu_i t . \qquad (6.3)$$

Substitute Eq. (6.2) and (6.3) into Eq. (6.1) and the result is

$$P = \alpha E = (\alpha_o + \frac{\beta A}{A} \sin 2\pi \nu_m t)(E_o \sin 2\pi \nu_i t) \qquad (6.4)$$

where ν_m is the molecular frequency of vibration. Expanding

$$P = \alpha_o E_o \sin(2\pi \nu_i t) + \beta E_o \sin(2\pi\nu_i t)\sin(2\pi\nu_m t)$$

$$= \alpha_o E_o \sin(2\pi \nu_i t) + \frac{\beta E_o}{2} [cos\ 2\pi(\nu_i - \nu_m)t - \cos 2\pi(\nu_i + \nu_m)t]. \quad (6.5)$$

Notice that this last equation looks like the equation for amplitude modulation, a process which is familiar to electrical engineers. We would therefore call the first term the "carrier"; the second and last terms are the "lower and upper sidebands", respectively. Remember, however, that we are discussing the results of the interaction of a time-varying electric field (light)* with a vibrating molecule! Historically, the first term in Eq. (6.5) is called the Rayleigh line. It represents the incident energy which is scattered with no change in frequency. The second term is the Stokes line, i.e., the portion of the incident energy scattered at a lower frequency. The last term is the anti-Stokes line, representing the portion of the incident energy which is scattered at a higher frequency. Polyatomic molecules exhibit more than one Stokes and one anti-Stokes line because such molecules have more than one normal mode of vibration.

The Raman lines, Stokes and anti-Stokes, do not have equal intensities as predicted by Eq. (6.5). It is observed experimentally that the Stokes lines are more intense than the anti-Stokes lines. The reason for the unequal intensities is the distribution as a function of temperature of the molecular population among the allowed energy states. The quantum theory which follows explains the appearance of Rayleigh, Stokes, and anti-Stokes lines as well as the intensity difference between Stokes and anti-Stokes lines.

The meaning of change in polarizability can be grasped by considering a classical model of a vibrating, linear, three-mass-spring system shown in Fig. 6.1a,b. The electric dipole vectors are p, the spring constant of the two identical springs is k, and the *external* applied sinusoidal electric field is $\underline{E}(t)$. In the absence of an external applied field, the molecule receives energy from the universe and energy will be absorbed if the energy is at or near a resonant frequency of the molecule. For this linear model there are 3N-5 normal modes of vibration, or 4 normal modes. Consider one of these modes which is the symmetric stretching mode, i.e., the center mass remains fixed and the two end masses vibrate in and out symmetrically with respect to the center.

To understand what happens to the symmetric vibration when the external field is applied, first let the field be a dc field with the system at rest. The charges will assume the positions shown in Fig. 6.1(a) with the left spring stretched to $(d + \Delta d)$ and the right spring compressed to $(d - \Delta d)$. Since there is only one molecule, the polarization $\underline{P}$ (electric dipole moment/unit volume) is equal to p. The polarization of this model before the field was applied is $p_L + p_r = de\underline{a}_x - de\underline{a}_x = 0$. The polarization with the dc field applied is

$$\underline{p}_L + \underline{p}_r = (d + \Delta d)\,\underline{a}_x - (d - \Delta d)\,e\underline{a}_x = 2\Delta de\underline{a}_x$$

---

*Light is a time-varying electromagnetic field, but we are concerned here with the electric vector only.

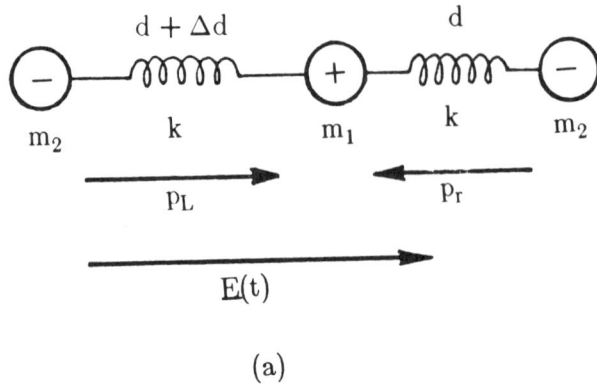

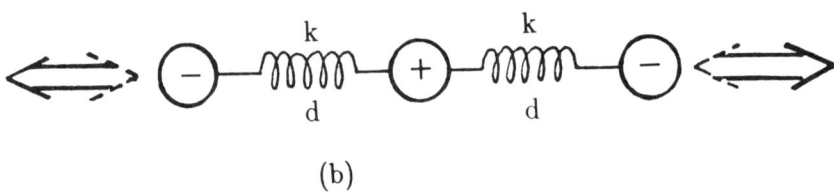

Figure 6.1 Vibration of a mass-spring system with and without an external applied electric field.

The polarizability change, $\Delta\alpha$, is therefore

$$\Delta\alpha = \frac{\Delta P}{E} = \frac{p_L + p_r}{E} = \frac{2\Delta d e}{E} a_x .$$

Now let E be time varying at a frequency 10 times the normal mode frequency (about the factor between visible frequencies and infrared frequencies). The Δd's will vary 10 times faster than the normal mode frequency, so the "forced" vibration frequency ν_L is "modulated" by the normal mode frequency ν_i. The resultant motion is a superposition of two forces, the external force due to $\underline{E} \cdot \underline{p}$ and restoring force of the spring. If the spring is non-linear, then overtones can also result.

One can visualize the covalent bonding between atoms in this molecule and the resultant perturbation of electronic energy states due to nuclear motion of vibration, i.e., the electronic states are vibration perturbed. The subsequent interaction of these states with an applied external field, coupled to the molecule through $\underline{E} \cdot \underline{p}$, results in the wave at light frequency ν_L interacting with perturbed electronic states, so the light is up- or down-shifted by the perturbation of frequency ν_i.

Different types of Raman scattering may occur depending on the frequency and intensity of the exciting laser light compared to electronic transition frequencies. Such types of Raman scattering are resonance Raman, stimulated Raman scattering, and surface enhanced Raman. These are specialized techniques that can have important engineering applications.

3 QUANTUM THEORY OF THE RAMAN EFFECT

Consider an incident photon of frequency ν_L from a He-Ne laser ($\lambda = 6328$Å). The incident photon has energy $h\nu_L$, the energy of the molecule from which the photon is scattered consists of translational energy, $\frac{mv^2}{2}$, and the internal energy, E_o, consisting of vibrational, rotational and electronic energy. Let the incident photon be scattered inelastically by the molecule such that the photon continues on with a different frequency ν'. The total molecular energy is modified as a result of the scattering. The energy of the system (molecule and radiation) is conserved, so

$$h\nu_L + \frac{mv_o^2}{2} + E_o = h\nu' + \frac{mv'^2}{2} + E'. \tag{6.6}$$

The momentum of a photon is the photon energy divided by the speed of light, so conservation of momentum in the system for a head-on collision is

$$\frac{h\nu_L}{c} + mv_o = \frac{-h\nu'}{c} + mv'. \tag{6.7}$$

The change in momentum is

$$\Delta(mv) = mv' - mv_o = \frac{h(\nu_L + \nu')}{c}. \tag{6.8}$$

In terms of *percent* change, the frequency ν' does not differ greatly from ν_L, so for the moment we assume $\nu_L \approx \nu'$ and write Eq. (6.8) as

$$\Delta(mv) = \frac{2h\nu_L}{c}. \tag{6.9}$$

The frequency for a Helium-Neon laser that is often used for Raman spectroscopy is $\nu_L = 4.7 \times 10^{14}$ sec^{-1}. Thus the value of $\Delta(mv)$ is

$$\Delta(mv) = \frac{2 \times 6.624 \times 10^{-27} \, erg. \, sec \times 4.7 \times 10^{14} \, sec^{-1}}{3 \times 10^{10} \, cm \, sec^{-1}}$$

$$\approx 2.0 \times 10^{-24} \, gm\text{-}cm/sec.$$

At room temperature, the momentum of the lightest atom, hydrogen, is 6.3×10^{-19} gm-cm/sec. Thus

$$\frac{mv_o^2}{2} \approx \frac{mv'^2}{2} \tag{6.10}$$

and so

$$h\nu_L + E_o \approx h\nu' + E' \tag{6.11}$$

or

$$\frac{E_o - E'}{h} = \nu' - \nu_L. \tag{6.12}$$

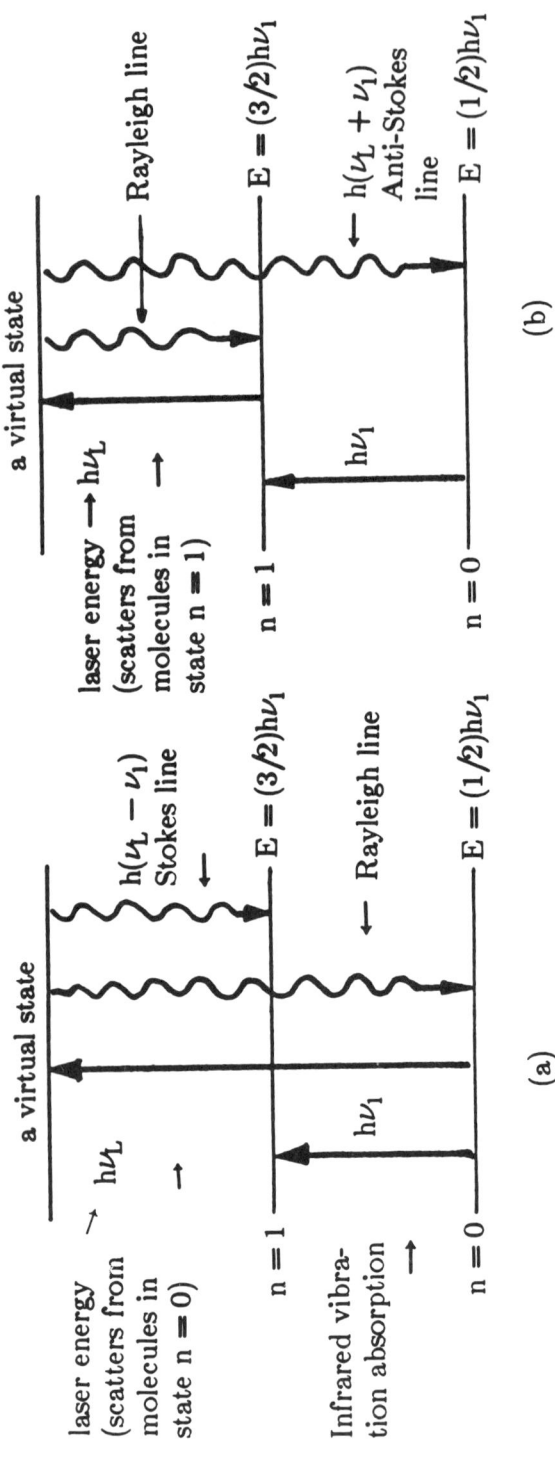

Figure 6.2 (a) Origin of a Stokes line, and (b) Origin of an anti-Stokes line.

The difference $E_o - E'$ is mostly vibrational energy. Consider the first two vibrational energy states of a molecule, $n = 0$ and $n = 1$. These states are shown in Fig. 6.2, and the absorption of a photon of infrared energy is designated as $h\nu_1$. An energetic photon of laser light, $h\nu_L$, interacts with a molecule in the state $n = 0$ and excites it to some much higher level which we call a virtual state. If then the molecule loses energy and decays back to $n = 0$, the original frequency ν_L is scattered and this scattered frequency is called the Rayleigh line. If the molecule decays back to $n = 1$ instead of $n = 0$, then the energy lost is $h(\nu_L - \nu_1)$. Notice that this energy differs from the original input energy, $h\nu_L$, by $h\nu_1$, or

$$h\nu_1 = h\nu_L - h(\nu_L - \nu_1). \tag{6.13}$$

Thus the scattered radiation contains a frequency $\nu_L - \nu_1$ which appears below the laser frequency by the difference ν_1. We measure the frequency displacement of the Stokes frequency from the exciting laser line to obtain the frequency ν_1 which corresponds to the *infrared* vibration transition $n = 0 \rightarrow 1$.

Now suppose that a photon of laser light, $h\nu_L$, interacts with a molecule in the state $n = 1$ and excites it to some other virtual state. If the molecule loses energy and decays back to $n = 1$, the original frequency ν_L is scattered and contributes to the Rayleigh line. If the molecule decays back to $n = 0$ the scattered energy is $h(\nu_L + \nu_1)$ and the frequency $\nu_L + \nu_1$ appears above the exciting frequency by the difference ν_1. Thus

$$h\nu_1 = h(\nu_L + \nu_1) - h\nu_1. \tag{6.14}$$

The relative intensities of the Stokes and anti-Stokes line is explained by considering the number distribution with temperature of the molecular population among the allowed energy states. This distribution is called a Boltzmann distribution which is

$$n_1 = n_o e^{-E/kT} \tag{6.15}$$

where n_o is the total number of molecules present, n is the number in the energy state E at the absolute temperature T, and k is Boltzmann's constant. Thus from Fig. 6.2, Eq. (6.15) becomes

$$n_1 = n_o e^{-(\frac{3}{2}h\nu_1 - \frac{1}{2}h\nu_1)/kT} = n_o e^{-h\nu_1/kT}. \tag{6.16}$$

In the first excited state at room temperature $h\nu_1 \gg kT$, so most molecules (about 99%) are in the vibrational state $n = 0$. Recall that the intensity of a spectral line depends upon the number of molecules which make that particular transition. Therefore we expect the Stokes line to be more intense because some of the incident laser beam scatters from molecules in the most populous state, $n = 0$.

One important application of Raman spectroscopy in electrical engineering is the evaluation of semiconductors. Raman gives information about dopant levels, induced strain, optimum composition of 3-4 semiconductor alloys, and the nature of the crystalline interface that effects properties of the overall device. The laser beam used to provide Raman excitation can be varied in frequency. Hence the depth of penetration of the light is varied and depth profiling of semiconductor characteristics can be accomplished.

CHAPTER

7

INTRODUCTION TO GROUP THEORY

1 REASONS FOR STUDYING METHODS OF GROUP THEORY

Some of the methods of group theory give powerful physical insight into a variety of problems which involve symmetry. The theory does so in an elegant, relatively uncomplicated manner. Many qualitative aspects of problems in molecular vibrations, semiconductors, crystal structure, and atomic energy states may be solved quickly with the aid of group theory, and quantitative problems in these areas may be simplified considerably. The theory offers a method of classifying molecules according to their symmetry, and complicated problems may be divided into smaller, less complicated parts. The methods are used by chemists and physicists for the interpretation of molecular spectra, i.e., to determine molecular structure, to identify modes of molecular vibration, etc. A perusal of engineering papers which discuss quantum electronic devices will reveal that group theory is used often in discussions of crystal and molecular structure.

Only a part of the theory of groups is necessary for applications to the problems which are discussed in this chapter, so we shall begin now to develop that portion of the theory of groups which is useful in solving and simplying problems of molecular structure and its relation to molecular vibration. Such knowledge is extremely useful in the design of molecular lasers because it is the molecular structure which determines the number of modes and types of modes.

2 DEFINITION OF AN OPERATION

The student is probably familiar with the term *operator* from calculus courses. The term is used to mean a rule by which a quantity or function is changed or transformed into another quantity or function. As two examples, one may perform operations on functions of variables or on geometrical structures. We shall be concerned primarily with operations on geometrical structures which represent molecular structures. The operations which are performed upon these structures are operations such as rotation, reflection and inversion.

An *operator* is denoted by a capital letter such as A, B, C, D,... or, in some cases, by capital letters with subscripts, such as A_1, A_2, B_1, B_2, etc. Thus, for example, the operator A may represent a rotation of a geometrical structure by 90° about some axis of symmetry, the operator B may mean reflection of that structure through one or more of its planes of symmetry, etc. Most often one uses *sets* of operators; a set contains the *elements* which are the operators. So A, B, C may be a set; or so may A, B, C, D; or A_1, B_1, etc.

Operators may be combined by certain rules to produce other operators. For example, the operator equation AB = C means that operation B is performed first, then operation A is performed upon that result to produce a resultant which is the same as if operation C were performed alone. The same operator may be applied successively, $AA = A^2$ or $AAA = A^3$, and so on.

3 DEFINITION OF A GROUP

Some particular set (collection) of operators may form what is termed a *group*, if its elements satisfy certain conditions. (The word "group" is used here in a strict mathematical sense, just as the word "field" in electromagnetic theory has a restricted definition). The conditions which the operators of a set must satisfy in order for the set to be a *group* are:
(a) The set is closed.
(b) The set contains the identity element, E.
(c) Every element of the set has an inverse.
(d) The associative law holds for all elements in the set, i.e., $CDF = (CD)F = C(DF)$.

Any set of elements (operators) which has a defined law of combination and for which (a), (b), (c), (d) are true is called a "group". Statement (a) above means that a combination of any two elements yields an element which is in the set. The letter E, the identity element, produces indistinguishable results when it operates upon another operator, for example, $EA = A$, $EB = B$, etc. The inverse of an operator, A^{-1}, means that $AA^{-1} = E = A^{-1}A$.

Sometimes the inverse of an operator may produce another operator of the same set.

4 EXAMPLES OF GROUPS—EXAMPLE 1

Consider the operator $\pm j = \pm \sqrt{-1}$ which is familiar to students of electrical engineering. This operator is used to shift phasor quantities by $\pm 90°$. Choose a set of four elements (operators), say A, B, C, E. Let

A = multiplication by j
B = multiplication by -1
C = multiplication by -j
E = multiplication by +1

and combine the operators by multiplication, combining all pairs of these operators in both orders, such as AB, BA, etc. The results of the combinations are

AA = B	BA = C	CA = E	EA = A
AB = C	BB = E	CB = A	EB = B
AC = E	BC = A	CC = B	EC = C
AE = A	BE = B	CE = C	EE = E

All possible pairs of these operators have been combined and you will note that the answers (right side of equation) are always elements in the *original* set A, B, C, D; so under this law of combination (multiplication) we always produce an element of the set. A neater way of representing these equations is by "multiplication table" shown below. The table has rows and columns like a matrix and we use the designation

Multiplication Table for Operators of First Example

A   B   =   C
(row)   (column)   (table)

	E	A	B	C
E	E	A	B	C
A	A	B	C	E
B	B	C	E	A
C	C	E	A	B

The operator E serves as the identity element, and the inverse of each operator $A^{-1} = C$, $B^{-1} = B$, $C^{-1} = A$, $E^{-1} = E$ produces only elements which are in the set. Thus it is correct to call this set of operators (A, B, C, E) a "group", because the definitions (a), (b), (c), (d) of a group are satisfied.

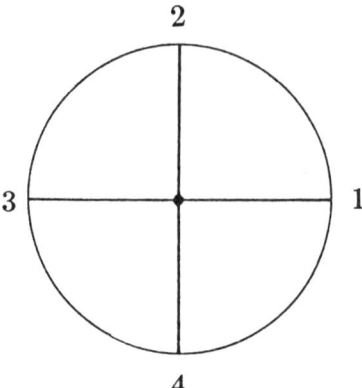

Figure 7.1 A 4-spoke wheel with fixed axis perpendicular to the page.

Example 2
Consider the clockwise axial rotations of a fixed-axis, 4-spoke wheel which produce a configuration indistinguishable from the original configuration. The operators are called *covering operations* and they have great significance in the study of molecular structure and molecular properties. The wheel is shown in Fig. 7.1. The operators for this example are defined as follows:

A = clockwise rotation by $\pi/2$ radians.
B = clockwise rotation by π radians.
C = clockwise rotation by $3\pi/2$ radians.
E = clockwise rotation by 2π radians. (identity)

Take two-operator combinations such as AA, AB, ...; BA, BB, ...; CA, CB, ...; EA, EB, ...; etc. Recall that CA, for example, means that operation A is performed first, then operation C is performed upon that result. The performance of these defined operations may be done now and the results are:

AA = B	BA = C	CA = E	EA = A
AB = C	BB = E	CB = A	EB = B
AC = E	BC = A	CC = B	EC = C
AE = A	BE = B	CE = C	EE = E

The multiplication table for this second example is shown below.

	E	A	B	C
E	E	A	B	C
A	A	B	C	E
B	B	C	E	A
C	C	E	A	B

Notice the permutations in the rows of this table. These permutations are what one would expect from a symmetrical structure like the wheel.

5 OTHER DEFINITIONS AND CONCEPTS

Now several other definitions need to be considered.
1. A group is said to be "finite" if it contains a finite number of elements.
2. The number of elements in a finite group is its "order", n.
3. A group is "Abelian" if all of its operators (elements) commute, that is if AB = BA, etc.
4. A group is "cyclic" if powers of some element yield all of the elements in the group.

Our first example of a group is finite, its order is $n = 4$, and it is Abelian and cyclic. The student should verify these facts. The group in our second example is finite, its order is $n = 4$, it is not Abelian because $AB \neq BA$ (are there other such inequalities?), and it is cyclic because $A^2 = B$, $A^3 = C$, $A^4 = E$.

In order to illustrate these concepts by a molecular model, consider a molecule of ammonia, NH_3, which has one nitrogen atom joined to three hydrogen atoms to form a pyramidal structure. This is an XY_3-type molecule, $X = N$, $Y = H$, which is shown below in Fig. 7.2. Much research has been done on the ammonia molecule both in the microwave and infrared regions of the spectrum, and this molecule was used as the first atomic frequency standard (ammonia maser). If the molecule is viewed from above, it appears planar as shown below in Fig. 7.3. Each axis of symmetry in the plane is shown by dotted lines and numbered 1, 2, 3. Also a symmetry axis, 4, passes through N perpendicular to the plane of the page. Consider axes 1, 2, 3 and 4 as fixed in space, but permit the molecule freedom to rotate as a rigid structure. Written below are operations which rotate the molecule with respect to these space-fixed axes and which bring the molecule to a configuration which is *indistinguishable* from the original configuration.

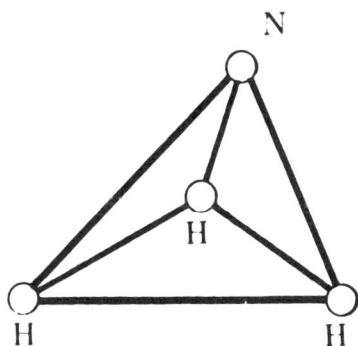

Fig. 7.2 The geometrical structure of ammonia, NH_3.

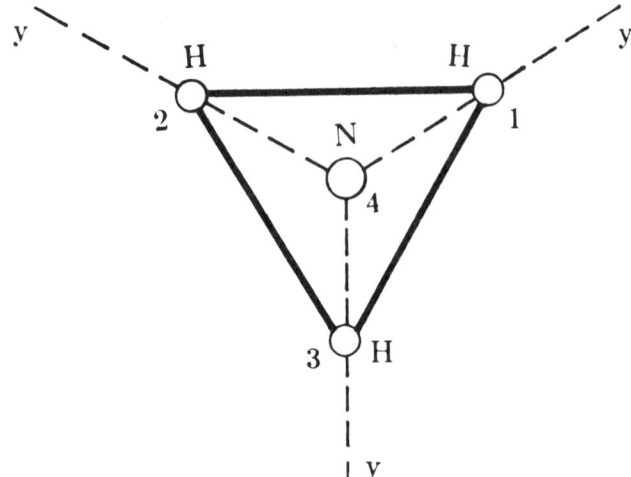

Figure 7.3 View of ammonia molecule from "above".

$$\begin{align}
E &= \text{identity} \\
A &= \text{clockwise rotation by } 120° \text{ about axis 4.} \\
B &= \text{clockwise rotation by } 240° \text{ about axis 4.} \\
C &= \text{rotation by } 180° \text{ about axis 1.} \\
D &= \text{rotation by } 180° \text{ about axis 2.} \\
F &= \text{rotation by } 180° \text{ about axis 3.}
\end{align}$$

Now take all possible binary multiplicative combinations of these operators as was done in the previous examples, and the following multiplication table is generated. The student should verify the table.

Multiplication Table for NH_3 Model

	E	A	B	C	D	F
E	E	A	B	C	D	F
A	A	B	E	D	F	C
B	B	E	A	F	C	D
C	C	D	F	E	A	B
D	D	F	C	B	E	A
F	F	C	D	A	B	E

Inverse

$$E^{-1} = E$$
$$A^{-1} = B$$
$$B^{-1} = A$$
$$C^{-1} = C$$
$$D^{-1} = D$$
$$F^{-1} = F$$

Notice that this group is finite, the order is 6, it is not Abelian and not cyclic. More will be said later about the use of molecular models and multiplication tables in the determination of molecular parameters, so keep these tables in mind.

Now four more important definitions from group theory are presented. These definitions are used extensively in relating between a molecular structure and the quantized frequencies of vibration for that structure.

6 SIMPLE ISOMORPHISM

Let there be two groups G and G' which have elements A, B, C, ... and A', B', C', ..., respectively. If $AB = C$ when $A'B' = C'$, i.e., a one-to-one correspondence between products in each group, then the groups G and G' are said to be *simply isomorphic*. Simple isomorphism is a very important property of groups, because groups which are simply isomorphic to each other will have one-to-one correspondence in their multiplication tables. This correspondence permits us to make important generalizations about similar geometrical configurations. It is shown later that it also permits one to use groups of *matrix operators* which are simply isomorphic to groups of geometrical operators in the elucidation of molecular vibrations.

7 ORDER, PERIOD

Any element of a group is said to have an "order", h, if for any element X, $X^h = E$, the identity. The value of h is understood to be the smallest integer which yields the identity. The period is the set $X, X^2, X^3, ..., X^n$ of any element X of a group and it is denoted by $\{X\}$. $\{X\}$ satisfies all of the group postulates and it forms a finite group called a cyclic group.

8 SUBGROUP

A subgroup is a finite group whose elements are contained in a larger group as an example, $\{A\} = \{B\} = E, A, B$ is a subgroup of the group of operators for NH_3. A subgroup is formed by taking the period of an element. In the NH_3 example, the periods are subgroups of the complete group. Note also that the order of each subgroup is a divisor of the order of the complete group.

9 CONJUGATE ELEMENTS

Given group elements X, A, B; if $B = XAX^{-1}$ then B is conjugate to A. The elements which are conjugate to A in example two are:

$$EAE^{-1} = A \qquad CAC^{-1} = CAC = CF = A$$
$$AAA^{-1} = AAB = AE = A \qquad DAD^{-1} = DAD = DC = A$$
$$BAB^{-1} = BAA = BB = A \qquad FAF^{-1} = FAF = FD = A$$

Thus A is conjugate to A. Conjugate elements have these properties:
(a) E is conjugate only to itself.
(b) Every element is conjugate to itself.
(c) If A is conjugate to B, then B is conjugate to A.
(d) If A is conjugate to B and C, then B and C are conjugate to each other.

In the following definition it will be seen that elements of a group which are conjugate to *each other* posses similar properties and may be placed together in a "class".

10 CLASSES OF ELEMENTS

In a group the set of elements which are conjugate to each other form a *class* of the group. There may be several classes in a group. A group can always be divided into classes, and no two classes have any common elements. The identity, E, is always in a class by itself. Classes are written as $C_1, C_2, C_3,...$. We see from example three that C_1 contains A and B; C_2 contains C, D, and F; and C_3 contains E. Notice that A and B are operations of the same kind, (rotations about one kind of axis) and that C, D, and F are operations of another kind (rotations about another kind of axis). Notice also that these three classes do not contain any common elements.

11 CATEGORIES OF OPERATIONS

We are ready now to consider two categories of operations which are used in the study of molecular vibrations. A choice between the two categories depends upon the information desired. These categories are:
(1) Covering operations—these operations are used in solving problems such as:
 (a) How do you use the natural symmetry of a molecule to advantage so that proper coordinates are chosen? (The proper choice of coordinates permits a complicated analytical problem to be broken up into smaller, simpler pieces).
 (b) What types of vibrational interaction can occur in a molecular structure?
 (c) How do you determine the number and the types of vibrational modes which occur in a given molecular structure?
 (d) What are the infrared and Raman selection rules for the fundamental modes of vibration? (What spectral lines will appear?)
 (e) What are the infrared and Raman selection rules for the overtone and combination modes of vibration?

(2) Permutation of identical atoms—permutation operations on identical atoms are used to calculate statistical weights of allowed energy states in molecules.

Our primary concern now is with operation (1), and these covering operations may be used several ways. Each way, however, produces a configuration of a molecule which is indistinguishable from its original configuration. The covering operations are:
(a) An operation which interchanges only identical atoms in a geometrical representation of a molecule, or
(b) An operation which carries a structure into itself, or
(c) An operation which carries a structure into a new configuration which is indistinguishable from the original configuration.

A covering operation may be a rotation about an axis of symmetry, a reflection in a plane of symmetry or a rotation about an axis of symmetry followed by a reflection in a plane which is perpendicular to that axis. Such a rotation followed by a reflection is called an *improper rotation*.

The symbols defined below will be used as a shorthand to describe the particular covering operations.

$C_k =$ rotation about an axis of symmetry by $2\pi/k$ radians, k an integer.

$S_k =$ improper rotation about an axis of symmetry by $2\pi/k$ radians, k and integer. (Improper rotation is a rotation, then a reflection in a plane perpendicular to the axis of rotation).

$I =$ inversion, which is an improper rotation by π radians. An inversion is covering operation which is equivalent to an improper rotation, S_2.

$\sigma_v, \sigma_h, \sigma_d =$ reflection in a vertical, horizontal, or dihedral plane, respectively. A dihedral plane bisects the angle between two horizontal symmetry axes which are 2-fold axes, i.e., C_2. Reflection in a plane of symmetry is equivalent to improper rotation through 0 radians.

If a molecular structure has more than one symmetry axis, the one of highest order (k) is termed the "principal symmetry axis" and it is chosen as the vertical z-axis in a coordinate system. Note that the intersection of two planes of symmetry is an axis of symmetry and, if the planes form an angle of π/k, the axis is k-fold. Furthermore, there exist (n-1) additional planes of symmetry separated by angles of π/k if a plane of symmetry contains a k-fold axis.

12 COORDINATE SYSTEMS FOR MOLECULES

We may describe the equilibrium position of the atoms in a symmetrical molecule by means of several body-fixed xyz coordinate systems which are equivalent. The center of gravity of the molecule is taken as the origin of the coordinate system, but otherwise there is more than one coordinate system possible to provide an equivalent description of the molecule. Consider the molecule $ClCH_3$ shown in Fig. 7.4. There is a three-fold axis of symmetry along the C-Cl bond and this axis is taken as the z-axis of an xyz right-hand coordinate system. Now imagine the molecule is viewed from above as shown in Fig. 7.5, where the z-axis is perpendicular to the page. There are three choices for the position of the y-axis, i.e., position 3 (as shown), position 2,

68 MOLECULES AND MOLECULAR LASERS FOR ELECTRICAL ENGINEERS

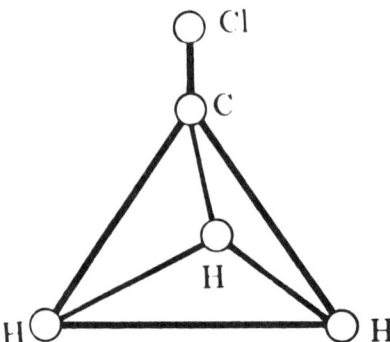

Figure 7.4 The molecule $ClCH_3$.

or position 1. For a left-hand coordinate system we have similar choices. The establishment of a coordinte system for a given molecule may proceed by choosing either of two equivalent viewpoints:

(1) The student may assume the individual atoms of a molecule are space-fixed and immovable, then choose different equivalent orientations of the chosen coordinate system by applying the covering operations to the coordinate system.
(2) The student may assume the molecular structure to be space-fixed and let the body-fixed coordinate system have only one special orientation, then interchange the positions of *identical* atoms in the system by covering operations on identical atoms.

The first viewpoint will be adopted for further discussion. An angle in a coordinate plane is positive if it turns x to y, y to z, or z to x.

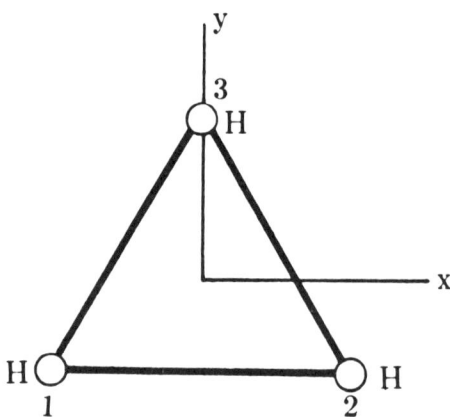

Fig. 7.5 The molecule $ClCH_3$ viewed from above.

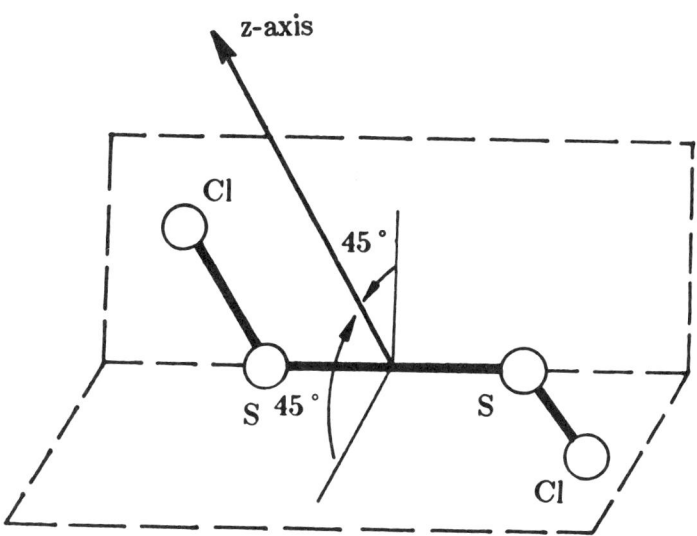

Figure 7.6 The molecule S_2Cl_2.

Let us illustrate these viewpoints with some molecules that have simple symmetry C_2 (S_2Cl_2, S_2Br_2, H_2O_2, S_2F_2). The molecule S_2Cl_2 is shown in Fig. 7.6 below. It has C_2 symmetry because it has only one element of symmetry other than the identity E, viz., a C_2 axis which is perpendicular to the line joining the sulfur atoms and which makes a 45° angle with each of the two planes shown. For this molecule we shall proceed to obtain the important items of information discussed earlier, i.e., the covering operations, the multiplication table, the inverse of each element, and the classes. The axis of highest symmetry is chosen as the z-axis of a right-hand coordinate system. Using viewpoint (1) we see that rotation of the coordinate system about the z-axis by π radians produces an indistinguishable configuration of the molecule. Thus for this molecule there are only two possible covering operations,

> E - the identity
> C_2 - rotation of π radians about the z-axis.

In this example the multiplication table is trivial, so we proceed to find the inverse of each element as discussed previously.

$$E^{-1} = E$$
$$C_2^{-1} = C_2$$

As discussed previously, the classes are determined by finding elements which are conjugate to each other.

$$EEE^{-1} = E$$
$$EC_2E^{-1} = C_2$$

$$C_2 E C_2^{-1} = E$$
$$C_2 C_2 C_2^{-1} = C_2$$

It is seen that C_2 and E are each in a class by themselves, so we write $C_1 = E$; (Class 2) $C_2 = C_2$. Notice again that each class contains an element which does something to the configuration which is quite unrelated to what the element in the other class does: thus the idea of separating operators into classes has a significant physical interpretation.

13 THE WATER MOLECULE

Let us go through the same process again for practice, but this time we shall use a molecular structure which is one rung up the ladder of structural complexity, water (H_2O), which has symmetry C_{2v}. The subscript v means that there are vertical planes of symmetry in addition to a 2-fold axis of symmetry. The molecule, it is symmetry axis, and its two planes of symmetry is shown in Fig. 7.7. The covering operations for H_2O are:

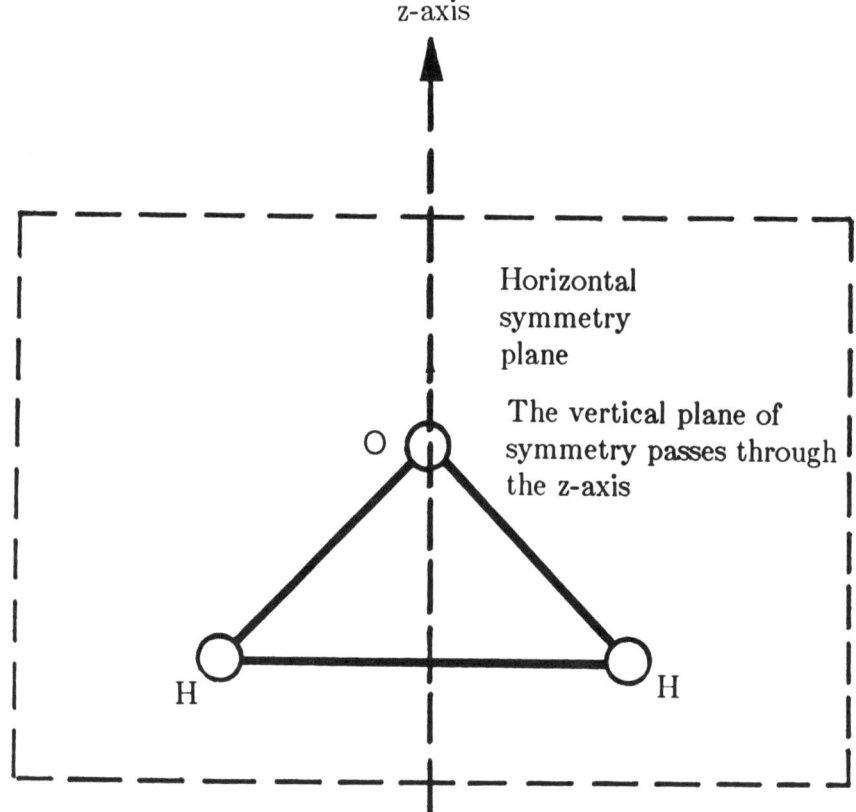

Figure 7.7 The water molecule, H_2O.

INTRODUCTION TO GROUP THEORY   71

$C_2 =$ rotation by π radians about the z-axis (recall that the z-axis is chosen as the axis of highest symmetry).
$\sigma_v =$ reflection in a vertical plane which contains the z-axis and the x-axis.
$\sigma_{v'} =$ reflection in a vertical plane which contains the z-axis and the y-axis.
$E =$ identity.

The inverse of each covering operator is:

$$C_2^{-1} = C_2$$

$$\sigma_v^{-1} = \sigma_v$$

$$\sigma_{v'}^{-1} = \sigma_{v'}$$

$$E^{-1} = E \ .$$

The multiplication table for H_2O is:

	E	C_2	σ_v	$\sigma_{v'}$
E	E	C_2	σ_v	$\sigma_{v'}$
C_2	C_2	E	$\sigma_{v'}$	σ_v
σ_v	σ_v	$\sigma_{v'}$	E	C_2
$\sigma_{v'}$	$\sigma_{v'}$	σ_v	C_2	E

Now we are ready to find the elements which are conjugate to each other. Recall we said B is conjugate to A if $B = XAX^{-1}$ for any element X in the group, so we begin by letting $A \equiv C_2$:

$$C_2 C_2 C_2^{-1} = C_2 E = C_2$$

$$\sigma_v C_2 \sigma_v^{-1} = \sigma_v \sigma_{v'} = C_2$$

$$\sigma_{v'} C_2 \sigma_{v'}^{-1} = \sigma_{v'} \sigma_v = C_2$$

$$E C_2 E^{-1} = E C_2 = C_2$$

Therefore C_2 is conjugate to itself only, and the first class is $C_1 = C_2$. Proceed to find the next class.

Let $A \equiv \sigma_v$:

$$C_2 \sigma_v C_2^{-1} = C_2 \sigma_{v'} = \sigma_v$$

$$\sigma_v \sigma_v \sigma_v^{-1} = \sigma_v E = \sigma_v$$

$$\sigma_{v'} \sigma_v \sigma_{v'}^{-1} = \sigma_{v'} C_2 = \sigma_v$$

$$E \sigma_v E^{-1} = E \sigma_v = \sigma_v$$

The second class is $C_2 = \sigma_v$, because σ_v is conjugate to itself only.

Let $A \equiv \sigma_{v'}$:

$$C_2 \sigma_{v'} C_2^{-1} = C_2 \sigma_v = \sigma_{v'}$$

$$\sigma_v \sigma_{v'} \sigma_v^{-1} = \sigma_v \sigma_v = \sigma_{v'}$$
$$\sigma_v \sigma_{v'} \sigma_{v'}^{-1} = \sigma_v E = \sigma_{v'}$$
$$E \sigma_v E^{-1} = E \sigma_{v'} = \sigma_{v'}$$

The third class is $C_3 = \sigma_{v'}$, because $\sigma_{v'}$ is conjugate to itself only.

Let $A \equiv E$:
$$C_2 E C_2^{-1} = C_2 C_2 = E$$
$$\sigma_v E \sigma_v^{-1} = \sigma_v \sigma_v^{-1} = E$$
$$\sigma_{v'} E \sigma_{v'}^{-1} = \sigma_{v'} \sigma_{v'}^{-1} = E$$
$$EEE^{-1} = E$$

The final class is $C_4 = E$ because E is conjugate to itself only. At this point if the student is uncertain about these operations, get a bag of gumdrops at a candy store and make up a molecular model using toothpicks. Such models are often helpful in visualizing in three dimensions the operations which are described by the multiplication table.

Note that if the water molecule were linear, as in the case of carbon dioxide (CO_2), there would be *one* plane of symmetry which contains all three atoms and one axis of symmetry. We designate CO_2 as having symmetry $D_{\infty v}$, i.e., any C_k and an infinite number of σ_v.

14 THE ETHYLENE MOLECULE

As a final example before moving on, let us consider the ethylene molecule, C_2H_4, which is shown in Fig. 7.8. This molecule has three 2-fold axes of symmetry with the z-axis chosen as shown and 2 vertical planes of symmetry. There is a horizontal plane of symmetry σ_h, and a center of symmetry at 0. Each axis serves as a 2-fold rotation-reflection axis, and each of the C_2's is also an S_2 (improper rotation by 180°). The right-hand coordinate system drawn to the right of the figure is placed at the center of symmetry with the orientation shown. This molecule is classified under a symmetry group called D_{2h}.

The covering operators for the ethylene molecule are defined as follows:
E = identity
A = rotation by π about the positive z axis.
B = rotation by π about the positive y axis.
C = rotation by π about the positive x axis.
D = improper rotation by π about the positive z axis.
F = improper rotation by π about the positive y axis.
G = improper rotation by π about the positive x axis.
H = reflection in a horizontal plane through the x-y axes.
I = reflection in a vertical plane through the y-z axes.

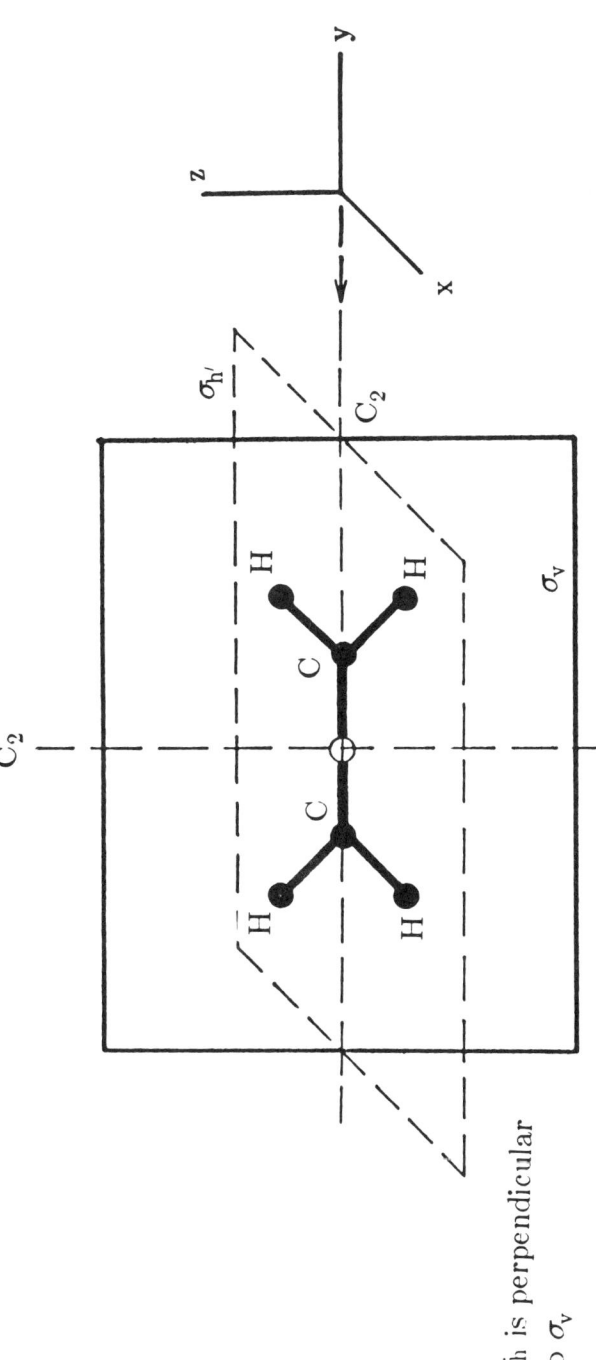

Figure 7.8 The molecule C_2H_4.

74 MOLECULES AND MOLECULAR LASERS FOR ELECTRICAL ENGINEERS

J = reflection in a vertical plane through the z-x axes.

The multiplication table for this group of order ten is obtained by the application of the covering operators to the righthand coordinate system shown in Fig. 7.9., assuming the individual atoms are space-fixed. We shall observe the effect of each covering operation upon the *original* orientation of the axis. The operations are illustrated now for clarity.

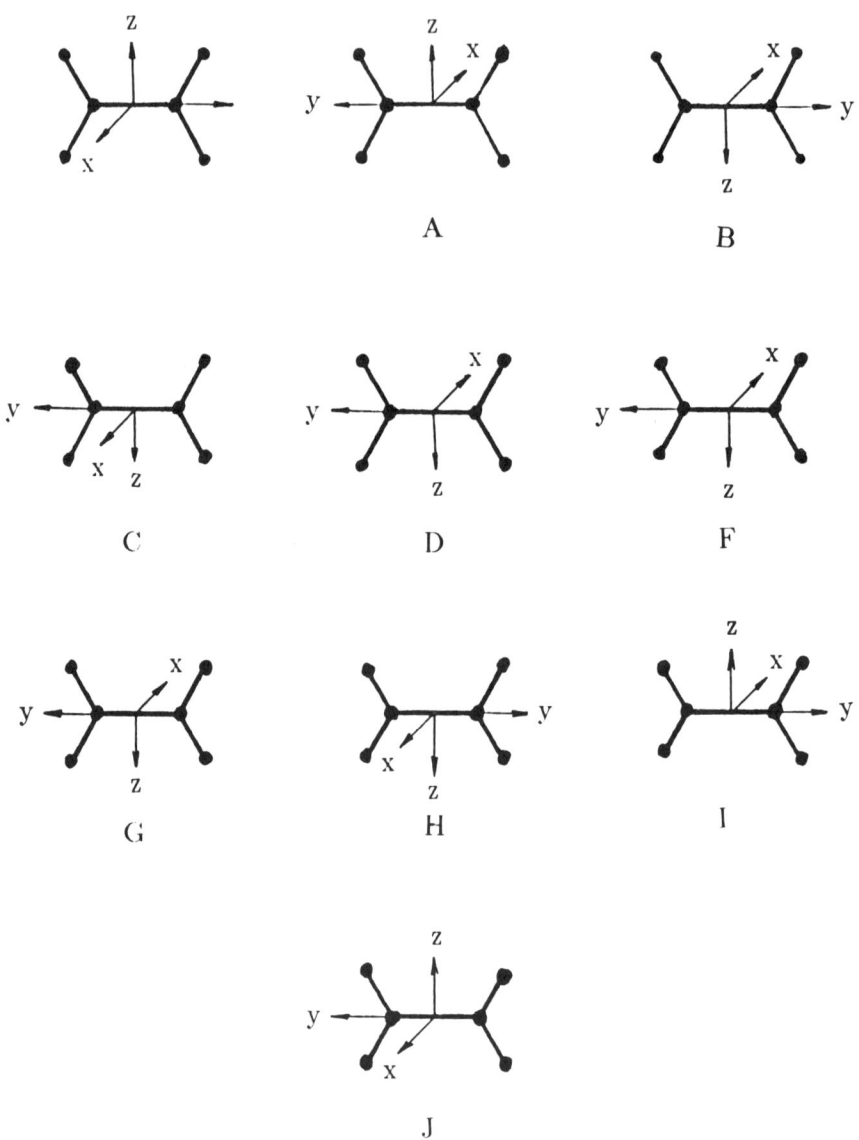

Figure 7.9 Effect of coverning operations on original orientation of the axis.

Note that covering operations F and G produce the same arrangement of the axes as does covering operation D. Therefore F and G are redundant and need not be included in the elements of the group, so the order of the required group of covering operations is now eight instead of ten. The calculation of the multiplication table should be done by the student for practice. The completed multiplication table for C_2H_4 is shown below.

	E	A	B	C	D	H	I	J
E	E	A	B	C	D	H	I	J
A	A	E	C	B	H	D	J	I
B	B	C	E	A	J	I	H	D
C	C	B	A	E	I	J	D	H
D	D	H	J	I	E	A	C	B
H	H	D	I	J	A	E	B	C
I	I	J	H	D	C	B	E	A
J	J	I	D	H	B	C	A	E

Each element is its own inverse, as may be seen from the multiplication table which has the identity down the diagonal line going from the top left to the bottom right of the table. It turns out again in this example that each element is in a class by itself. Verification of this fact is left as another exercise. Lest the reader conclude that all elements always end up in classes by themselves, let me assure you that they do not. We shall proceed now to determine the effect of covering operations on *coordinates*.

15 THE EFFECT OF COVERING OPERATIONS ON COORDINATES

Let us consider first the rotation of a right-hand xyz coordinate system about the z-axis. Such rotations should be familiar to the student from studies in electromagnetic theory, analytic geometry, and dynamics. Consider a point P in Fig. 7.10 which has coordinates x, y, z or x', y', z'. Either the primed or the unprimed set of coordinates is a description of point P with respect to the origin O. The geometrical operation by which one obtains the primed set of axes from the unprimed set is simply the rotation about O of the unprimed coordinate system through an angle θ to a new position represented by the primed coordinates. We call this operation $C(\theta)$. The rotation is described as follows:

$$x'_n = \sum_{j=1}^{3} \alpha_{jn} x_j \qquad (7.1)$$

where $x'_1 = x'$, $x'_2 = y'$, $x'_3 = z'$, the α_{jn} are the direction cosines which describe the rotation, and $x_1 = x$, $x_2 = y$, $x_3 = z$. Equation (7.1) may be rewritten also as

$$x_j = \sum_{n=1}^{3} \alpha_{nj} x'_n \qquad (7.2)$$

which expresses the unprimed coordinates in terms of the primed ones. Thus in an xyz

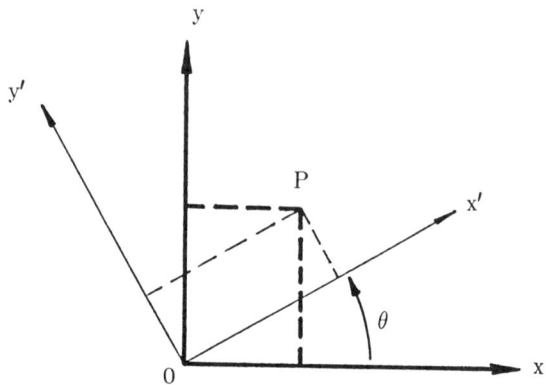

Figure 7.10 Rotation of x-y axes through the angle θ.

system Equations (7.1) and (7.2) become

$$x = x' \cos \theta - y' \sin \theta$$
$$y = x' \sin \theta + y' \cos \theta$$
$$z = z'$$

or

$$x' = x \cos \theta + y \sin \theta$$
$$y' = -x \sin \theta + y \cos \theta$$
$$z' = z \ .$$

These simple geometrical relations are the basis for the determination of the effects of *all* the covering operations to be discussed.

Consider next an improper rotation through an angle θ about the z-axis. Recall that an improper rotation is a rotation followed by a reflection. To perform the improper rotation the unprimed coordinate system is rotated by an angle θ around the z-axis and then the z-axis is inverted. This improper rotation is shown in Fig. 7.11.

The relationships among the coordinates of the unprimed and primed systems are

$$x = x'\cos \theta - y' \sin \theta \qquad x' = x \cos \theta + y \sin \theta$$
$$y = x' \sin \theta + y' \sin \theta \qquad y' = -x \sin \theta + y \cos \theta$$
$$z = -z' \qquad z' = -z$$

This covering operation is represented by $S(\theta)$.

The inversion of coordinates is simply an improper rotation through 180°. The old and new coordinates are related by

$$x = -x', y = -y', z = -z' \ .$$

INTRODUCTION TO GROUP THEORY 77

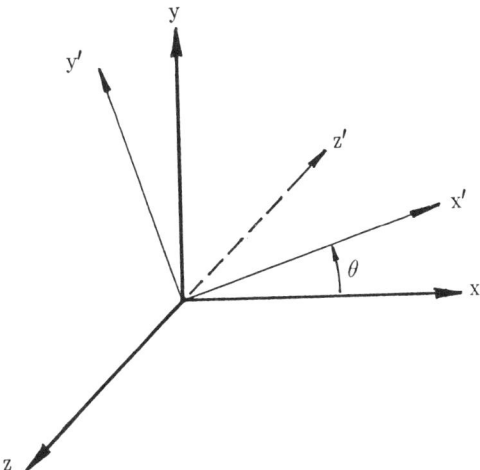

Figure 7.11 Improper rotation through an angle θ about the z-axis.

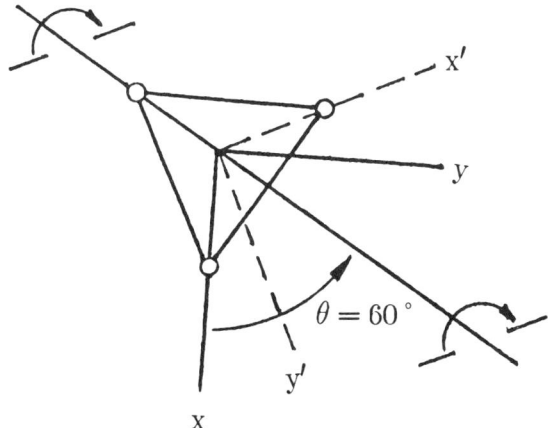

Figure 7.12 Covering operation $C_2(\theta)$ for the XY_3 molecule.

For the planar XY_3 molecule we define a covering operation $C_2(\theta)$, proper rotation by π radians about a two-fold axis in the xy plane and making an angle θ with the original x axis. The operation is shown in Fig. 7.12 for the XY_3 molecule and the relationships among the coordinates for this covering operation are

$$x = x'\cos 2\theta + y'\sin 2\theta \qquad x' = x\cos 2\theta + y\sin 2\theta$$
$$y = x'\sin 2\theta - y'\cos 2\theta \qquad y' = x\sin 2\theta - y\cos 2\theta$$
$$z = -z' \qquad z' = -z$$

The next covering operation to consider is reflection in a plane of symmetry. To be as general as possible consider a vertical plane of symmetry and let the intersection of this plane with the xy-plane make an angle θ with the original x-axis. This situation

78 MOLECULES AND MOLECULAR LASERS FOR ELECTRICAL ENGINEERS

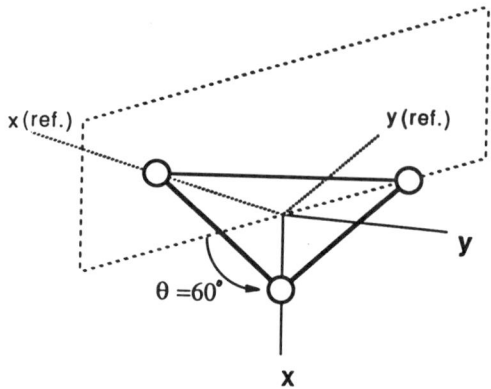

Figure 7.13 Reflection in a plane of symmetry for the XY_3 molecule. The reflected x and y axes are shown (mirror images).

is shown in Fig. 7.13 for the XY_3 molecule, $\theta = 60°$. We wish to find operations which are equivalent to reflection, so that we may write down a set of transformation equations. These operations are rotation of the original xy-axes through an angle of 2θ about the z-axis followed by an inversion of the *new* y-axis position (labeled y' in Fig. 7.11). These operations yield x(reflected) and y(reflected), and so they are *equivalent* to reflection. The old and new coordinates are

$$x = x' \cos 2\theta + y' \sin 2\theta \qquad x' = x \cos 2\theta + y \sin 2\theta$$
$$y = x' \sin 2\theta - y' \cos 2\theta \qquad y' = x \sin 2\theta - y \cos 2\theta$$
$$z = z' \qquad z' = z$$

This covering operation is represented by $\sigma_v(\theta)$.

If a square array is constructed from the coefficients of the variables for each set of transformation equations, there are formed "transformation matrices" (one matrix for each operation) which may be treated as *operators*. These matrix operators form *groups* which are simply isomorphic to the geometrical operator groups, and the matrix operators may also be put into classes, just as we did for the geometrical operators from which these matrix operators are derived. All of the matrix operators in a class share an important common property—*the sum of the diagonal elements is the same for all of the matrix operators in that class*. This sum is called the *character* because it characterizes all of the operators in a class. The characters of a group are *very important* in the application of group theory to physical problems.

The transformation matrix which corresponds to a geometrical covering operation is used as a *representation* of that covering operation. Therefore the geometrical operators are replaced by their corresponding matrix operators. To summarize our results, the matrix representations of the geometrical covering operations are listed. $C'_2(\theta)$ is proper rotation by π about a 2-fold axis in the x-y plane and making an angle θ with the original x axis. $C_2(\theta)$ is proper rotation through an angle θ about the z-axis. $\sigma_v(\theta)$ is reflection in the vertical plane whose intersection with the x-y plane makes an angle θ with the original x-axis. The arrow indicates that the operator is replaced by the matrix operator.

$$E \to \begin{bmatrix} 1 & 0 & 0 \\ 0 & 1 & 0 \\ 0 & 0 & 1 \end{bmatrix} \qquad C(\theta) \to \begin{bmatrix} \cos\theta & \sin\theta & 0 \\ -\sin\theta & \cos\theta & 0 \\ 0 & 0 & 1 \end{bmatrix}$$

$$S(\theta) \to \begin{bmatrix} \cos\theta & \sin\theta & 0 \\ -\sin\theta & \cos\theta & 0 \\ 0 & 0 & -1 \end{bmatrix} \qquad I \to \begin{bmatrix} -1 & 0 & 0 \\ 0 & -1 & 0 \\ 0 & 0 & -1 \end{bmatrix}$$

$$C'_2(\theta) \to \begin{bmatrix} \cos 2\theta & \sin 2\theta & 0 \\ \sin 2\theta & -\cos 2\theta & 0 \\ 0 & 0 & -1 \end{bmatrix} \qquad \sigma_v(\theta) \to \begin{bmatrix} \cos 2\theta & \sin 2\theta & 0 \\ \sin 2\theta & -\cos 2\theta & 0 \\ 0 & 0 & 1 \end{bmatrix}$$

We must now find the inverse of these matrix representations, just as we found the inverses of the geometrical covering operators. The definition of the inverse, A^{-1}, of a matrix A, is that matrix for which $AA^{-1} = A^{-1}A = E$. The inverse of the matrix representations are listed here. The derivations are left as an exercise for the student.

$$E^{-1} \to \begin{bmatrix} 1 & 0 & 0 \\ 0 & 1 & 0 \\ 0 & 0 & 1 \end{bmatrix} \qquad C^{-1}(\theta) \to \begin{bmatrix} \cos\theta & -\sin\theta & 0 \\ \sin\theta & \cos\theta & 0 \\ 0 & 0 & 1 \end{bmatrix}$$

$$S^{-1}(\theta) \to \begin{bmatrix} \cos\theta & -\sin\theta & 0 \\ \sin\theta & \cos\theta & 0 \\ 0 & 0 & -1 \end{bmatrix} \qquad I^{-1} \to \begin{bmatrix} -1 & 0 & 0 \\ 0 & -1 & 0 \\ 0 & 0 & -1 \end{bmatrix}$$

$$C'_2(\theta)^{-1} \to \begin{bmatrix} \cos 2\theta & \sin 2\theta & 0 \\ \sin 2\theta & -\cos 2\theta & 0 \\ 0 & 0 & -1 \end{bmatrix} \qquad \sigma_v^{-1}(\theta) \to \begin{bmatrix} \cos 2\theta & \sin 2\theta & 0 \\ \sin 2\theta & -\cos 2\theta & 0 \\ 0 & 0 & 1 \end{bmatrix}$$

Note that $E = E^{-1}$, $I = I^{-1}$, $C'_2(\theta) = C'_2(\theta)^{-1}$ and $\sigma_v(\theta) = \sigma_v^{-1}(\theta)$.

Problem: Return to the previous examples of H_2O and C_2H_4, and show that multiplication of these matrices satisfy the multiplication table for the corresponding geometrical operators.

Also, we shall refer to the examples of S_2Cl_2 and H_2O in order to write down by inspection the transformation matrices (representations) for these molecules. For S_2Cl_2 the representations are:

$$E = \begin{bmatrix} 1 & 0 & 0 \\ 0 & 1 & 0 \\ 0 & 0 & 1 \end{bmatrix}; \quad C_2 \to C(\pi) = \begin{bmatrix} \cos\pi & \sin\pi & 0 \\ -\sin\pi & \cos\pi & 0 \\ 0 & 0 & 1 \end{bmatrix} = \begin{bmatrix} -1 & 0 & 0 \\ 0 & -1 & 0 \\ 0 & 0 & 1 \end{bmatrix}$$

For H_2O the representations are:

$$E = \begin{bmatrix} 1 & 0 & 0 \\ 0 & 1 & 0 \\ 0 & 0 & 1 \end{bmatrix}; \quad C_2 \to C(\pi) = \begin{bmatrix} -1 & 0 & 0 \\ 0 & -1 & 0 \\ 0 & 0 & 1 \end{bmatrix};$$

$$\sigma_v \to \sigma_v(0) = \begin{bmatrix} \cos 0 & \sin 0 & 0 \\ \sin 0 & -\cos 0 & 0 \\ 0 & 0 & 1 \end{bmatrix} = \begin{bmatrix} 1 & 0 & 0 \\ 0 & -1 & 0 \\ 0 & 0 & 1 \end{bmatrix}$$

$$\sigma'_v \to \sigma_v(\pi/2) = \begin{bmatrix} \cos \pi & \sin \pi & 0 \\ \sin \pi & -\cos \pi & 0 \\ 0 & 0 & 1 \end{bmatrix} = \begin{bmatrix} -1 & 0 & 0 \\ 0 & 1 & 0 \\ 0 & 0 & 1 \end{bmatrix}$$

16 SPECIAL PROPERTIES AND REPRESENTATIONS

Not all arrays of numbers have such interesting properties as these arrays. If arrays do have certain special properties, then operations can be performed on the arrays which have important physical significance in regard to both molecular structure and modes of molecular vibration.

Let us choose the matrix $C(\theta)$ and interchange its rows and columns; at the same time take the complex conjugate of each element in the matrix. We have just formed the "adjoint" matrix by this process, written $C(\theta)^+$. Notice that in so doing, the resulting matrix, $C(\theta)^+$, is the same as $C^{-1}(\theta)$, i.e., $C(\theta)^+ = C^{-1}(\theta)$. If we repeat this process for each matrix representation, we see that in each case the adjoint matrix is equal to the inverse. The matrix A is *unitary* if $A^+ = A^{-1}$. Thus all of these representations are unitary! The determinant of a unitary matrix is equal to one, and of special importance is the following: *the trace (sum of the diagonal elements) of a matrix B is unchanged by a transformation of the type ABA^{-1}, A unitary*. This transformation is called a *unitary transformation*, and such transformations are important in applications of group theory to molecular problems, quantum mechanics, and solid-state theory.

Notice also that all of the matrix representations for S_2Cl_2 and H_2O are similar in one respect—they are all *step* matrices. A step matrix is a diagonal matrix which contains smaller square matrices for the nonzero elements. The representation $C(\pi)$ for H_2O is

$$\text{2-fold step} \to \begin{bmatrix} -1 & 0 & | & 0 \\ 0 & -1 & | & 0 \\ -- & -- & -|- & -- \\ 0 & 0 & | & 1 \end{bmatrix} \leftarrow \text{one-fold step}$$

and the "steps" are partitioned by the dashed lines. When the matrices are partitioned by steps, the steps are a representation of the group of operators (as are the matrices themselves). The representation $C(\pi)$ for S_2Cl_2 may be partitioned into three one-fold steps as shown.

$$\begin{bmatrix} -1 & | & 0 & | & 0 \\ -- & -|- & -- & -|- & -- \\ 0 & | & -1 & | & 0 \\ -- & -|- & -- & -|- & -- \\ 0 & | & 0 & | & 1 \end{bmatrix}$$

and *each step is a representation of the group*. The steps which are present in classes E and C_2 for S_2Cl_2 are written below.

Representations (Steps) by Classes for S_2Cl_2

	E	C_2			E	C_2
R_1	1	1		R_1	1	1
R_2	1	-1	$\longrightarrow$	R_2	1	-1
R_3	1	-1				

On the left, representations R_2 and R_3 are equivalent, so the table is shortened to the form shown at the right.

A similar representation is shown for H_2O. From inspection of the matrix representions for this molecule, the steps are easily seen. They are

$$E: \quad 1, \quad 1, \quad 1$$
$$C_2: \quad -1, \quad -1, \quad 1$$
$$\sigma_v: \quad 1, \quad -1, \quad 1$$
$$\sigma'_v: \quad -1, \quad 1, \quad 1.$$

The representations by classes for H_2O are

Classes

	E	C_2	σ_v	σ_v'
R_1	1	-1	1	-1
R_2	1	-1	-1	1
R_3	1	1	1	1

Notice that no two representations are the same.

In general, the matrix A is "reducible" when a unitary transformation BAB^{-1} changes A into a matrix with more steps than it had originally. For example, a 3×3 matrix is reducible if a unitary transformation produces a 2-fold step and a single step (or if the transformation produces three single steps). Thus the representations *of a group* are reducible if all the matrices in that representation are reducible in corresponding ways.

The critical reader may have noticed that the table of representations for S_2Cl_2 is square, but the table of representations for H_2O is not square. There are always as many linearly independent irreducible representations of a group as there are classes of elements. Since there are four classes of elements for H_2O, there exists one more representation and we must find it. There are two methods of finding such a remaining representation:
(a) We may use the mathematical definition of linear independence and solve simultaneous equations for the desired representations.

82 MOLECULES AND MOLECULAR LASERS FOR ELECTRICAL ENGINEERS

(b)   We may look for *another* group which is simply isomorphic to the original group. If another group is found, it will contain some representation(s) which have steps that differ from the steps we already used, and we may use these *different* steps to complete our table of representations.

Without considerable experience and practice (b) is the more difficult of the two methods, so method (a) is applied below to find the one remaining representation for H_2O. We write:

[Number of irreducible representations = number of classes].

The irreducible representations form orthogonal vectors so

$$R_1 \cdot R_4 = R_2 \cdot R_4 = R_3 \cdot R_4 = 0 \quad (7.3)$$

where R_4 is the one remaining representation to be found.

$$R_1 = (1, -1, 1, -1)$$
$$R_2 = (1, -1, -1, 1)$$
$$R_3 = (1, 1, 1, 1)$$
$$R_4 = (a, b, c, d)$$
$$R_1 R_4 = a - b + c - d = 0$$
$$R_2 R_4 = a - b - c + d = 0$$
$$R_3 R_4 = a + b + c + d = 0$$

The simultaneous solution yields $a = b = -c = -d$. If we choose $a = b = 1$, then $c = d = -1$, so

$$R_4 = (1, 1, -1, -1).$$

A check for orthogonality shows that Eq. (7.3) is satisfied. This fourth representation is now included in the representation by classes for H_2O and the four representations for H_2O are

	E	C_2	σ_v	σ'_v
R_1	1	-1	1	-1
R_2	1	-1	-1	1
R_3	1	1	1	1
R_4	1	1	-1	-1

In both examples, S_2Cl_2 and H_2O, the original representations R_1, R_2, R_3 resulted from matrix representations which were reduced into one-fold steps. For other types of molecules a matrix representation may reduce to a two-fold step and a one-fold step. Such is the case for ammonia, NH_3, and its table of representations is shown below. Notice that R_1 and R_2 contain one-fold steps, whereas R_3 contains two-fold steps.

INTRODUCTION TO GROUP THEORY 83

Representations by Classes for NH_3

Class →	C_1	C_2		C_3		
Operator →	E	A	B	C	D	F
R_1	1	1	1	1	1	1
R_2	1	1	1	-1	-1	-1
R_3	$\begin{bmatrix} 1 & 0 \\ 0 & 1 \end{bmatrix}$	$\begin{bmatrix} -\frac{1}{2} & \frac{\sqrt{3}}{2} \\ -\frac{\sqrt{3}}{2} & \frac{1}{2} \end{bmatrix}$		$\begin{bmatrix} 1 & 0 \\ 0 & 1 \end{bmatrix}$	$\begin{bmatrix} \frac{1}{2} & \frac{\sqrt{3}}{2} \\ \frac{\sqrt{3}}{2} & -\frac{1}{2} \end{bmatrix}$	$\begin{bmatrix} \frac{1}{2} & -\frac{\sqrt{3}}{2} \\ \frac{\sqrt{3}}{2} & -\frac{1}{2} \end{bmatrix}$

17 USEFUL PROPERTIES OF REPRESENTATIONS BY CLASSES; THE CHARACTER TABLE

Representations by classes are used to obtain a *table of characters*, our ultimate goal in this section. *Given a character table, many important molecular properties can be easily calculated.* Recall that a "character" of a matrix is the sum of its diagonal elements. The table of characters is always square, and for the examples S_2Cl_2 and H_2O, we already have the character table, because the step matrices are one-fold. For NH_3, representations R_1 and R_1 are one-fold matrices and so the characters are the numbers in the first two lines. For R_3, the sum of the diagonal elements under each class is 2, -1, -1, 0, 0, 0.

The characters for NH_3 under each class are:

	C_1	C_2		C_3		
	E	A	B	C	D	F
R_1	1	1	1	1	1	1
R_2	1	1	1	-1	-1	-1
R_3	2	-1	-1	0	0	0

Notice that the characters for R_1 under classes C_2 and C_3 are redundant. A similar redundancy under classes C_2 and C_3 occurs for R_2 and R_3. Thus the set of characters under a class "characterize" that class because each operator in a class has the *same* character, but these redundancies yield no additional information. The character table for NH_3 (XY_3 molecule) is rewritten to eliminate these redundancies and becomes

	C_1	C_2	C_3
R_1	1	1	1
R_2	1	1	-1
R_3	2	-1	0

The character tables for S_2Cl_2 and H_2O are the same as their tables of representations because each operator for these molecules is in a class by itself and all of the reduced representations are one-fold steps.

The following are important properties of characters:
(a) characters of different representations form orthogonal vectors.
(b) the sum of squares of characters in a representation equals the order of the group.
(c) elements in a class have the same characters.
(d) any character of the identity operator E is the order of that representation containing the character.
(e) the character of a matrix is unchanged by a unitary transformation upon that matrix.

18 SUMMARY

The steps which have been taken in this chapter to arrive at the character table are listed for the convenience of the student.
(a) A geometrical model of the molecule is chosen.
(b) The covering operations are selected and executed, then a multiplication table is constructed.
(c) The inverse of each operator is obtained from inspection of the multiplication table.
(d) The classes are determined.
(e) Matrix operators are chosen which are simply isomorphic to the covering operators. Numerical values of the matrix elements are obtained by substitution of the correct values of the angles θ. Values of θ depend upon the particular molecule in question.
(f) These matrix representations of the original group of covering operators are each reduced into smaller steps.
(g) The table of representations is constructed, keeping in mind that there must be as many representations as there are classes.
(h) The characters of the representations are determined.
(i) Redundancies, if any, are eliminated.

This procedure is not time consuming, once a little practice is acquired, and remember one very important fact—*one* character table serves for *all* molecules having the same symmetry. This fact alone shows the power of group theoretic analysis.

19 SPECIES

The irreducible representations will be given new symbols now because the types of fundamental modes which occur in molecules are classified most easily according to their symmetry under the covering operations. Therefore irreducible representations are now *renamed* as "species". The species are listed below with their symmetry characteristics.

        A - symmetrical (character = +1) under C_2.
        B - antisymmetrical (character = -1) under C_2.
        1 (subscript) - symmetrical under σ_v
        2 (subscript) - antisymmetrical under σ_v

Vibration types that are symmetric or antisymmetric to σ_h are designated by a prime or a double prime (superscript), respectively. A species is just another description of an irreducible representation, so the representations R_1 and R_2 in the character table for S_2Cl_2 may be replaced by A and B, respectively; the representations $R_1, R_2, R_3,$ and R_4 in the character table for H_2O may be replaced by A_1, A_2, B_1, B_2, respectively, etc. Other classifications are:

    E - doubly degenerate vibration (the irreducible representation is a 2×2 matrix)
    F - triply degenerate vibration (the irreducible representation is a 3×3 matrix)
    g - symmetric with respect to a center of symmetry
    u - antisymmetric with respect to a center of symmetry

For linear molecules (point groups $D_{\infty v}$ and $C_{\infty v}$) the classifications are those used for electronic states of homonuclear diatomic molecules, viz.,

    Σ^+ - symmetric with respect to a plane of symmetry through the molecular axis
    Σ^- - antisymmetric with respect to a plane of symmetry through the molecular axis

π, Δ, ϕ - degenerate vibrations with degeneracies of 2, 3, and 4, respectively.

CHAPTER

8

USES OF THE CHARACTER TABLE

1 INTRODUCTION

A great deal of information can be obtained about a system which has some symmetry by the study of its transformation properties. Physical quantities whose transformation properties are of interest in the study of molecules are quantities such as

(a) electric dipole moment and its components
(b) polarization, polarizability, and their components
(c) linear displacement
(d) normal coordinates
(e) angular momentum and its components
(f) eigenfunctions of various operators

These quantities transform like one or more of the irreducible representations in a character table: because they do, these quantities can be categorized in a systematic way and the important results applied to a variety of point groups. It is here that we begin to see the elegant simplicity of the group theoretic approach to the solution of complicated problems. We will need to construct a *character for each physical quantity* of interest—from then on, it is a matter of doing some arithmetic.

2 TRANSFORMATIONS OF CHARACTERS UNDER SAMPLE OPERATIONS

We begin by learning how the character of a given physical quantity transforms under a sample operation from each class. This information is needed to learn the "classification" of the character, i.e., in a given class, what representations in the character table add algebraically to give the character of the physical quantity.

To illustrate, the electric dipole moment is chosen because it transforms like the coordinates. Recall that the definition of electric dipole moment is charge times a displacement of that charge from some origin. The charge is a scalar quantity but the displacement is a vector quantity. The dipole moment is written as

$$\mathbf{p} = \mathbf{i}\, p_x + \mathbf{j}\, p_y + \mathbf{k}\, p_z,$$

where $\mathbf{i}, \mathbf{j},$ and $\mathbf{k}$ are unit vectors in the x, y, z direction, respectively. The components $p_x, p_y,$ and p_z of the dipole moment for a collection of charged particles, like those in a molecule, are

$$p_x = \sum_i q_i x_i,\ p_y = \sum_i q_i y_i,\ p_z = \sum_i q_i z_i. \tag{8.1}$$

Let us consider the matrix group which represents S_2Cl_2 and compute the character of the electric dipole moment for this group. The classes were determined earlier as $C_1 = E, C_2 = C(\theta)$.

$$C_1 = E \qquad\qquad C_2 = C(\theta)$$

$$\begin{Vmatrix} 1 & 0 & 0 \\ 0 & 1 & 0 \\ 0 & 0 & 1 \end{Vmatrix} \qquad \begin{Vmatrix} \cos\theta & \sin\theta & 0 \\ -\sin\theta & \cos\theta & 0 \\ 0 & 0 & 1 \end{Vmatrix}$$

The character of the dipole moment, χ_m, for each class is $\chi_m(E) = 3$, $\chi_m(c(\theta)) = 1 + 2\cos\theta$ (adding diagonal elements). For this molecule, $\theta = \pi$, so $\chi_m(C_m \pi) = 1 + (-2) = -1$. For comparison, these two characters are written below the character table for S_2Cl_2 (see Table 8.1).

Table 8.1 Dipole moment characters χ_{mp} for S_2Cl_2.

	E	C_2
R_1	1	1
R_2	1	-1
χ_{mp}	3	-1

Under each class, the character of the physical quantity, in this case $\chi_{m\rho}$, will always be some linear combination of the representations in that class.

$$\chi_{m\rho} = k_1\chi_{1\rho} + k_2\chi_{2\rho}, \qquad (8.2)$$

where k_1 and k_2 are integral numbers, and $\chi_{1\rho}$, $\chi_{2\rho}$ are the characters under the class ρ in irreducible representations R_1 and R_2. This equation says that $\chi_{m\rho}$ is a *reducible* quantity. Sometimes the constants are easily deduced by inspection of the character table, but in group theory there is a systematic method for calculating the k's from the *expansion theorem*. The expansion theorem states that under a class any reducible representation (such as $\chi_{m\rho}$) can be written in terms of the irreducible representations of the group. Thus we write

$$\chi_{m\rho} = \sum_j k_j \chi_{j\rho} \qquad (8.3)$$

where $x_{j\rho}$ is the character under class ρ in the jth irreducible representation. The integral numbers k_j are given by

$$k_j = \frac{1}{n} \sum_\rho g_\rho \, \chi_{j\rho} \chi_\rho \qquad (8.4)$$

where the sum goes over the different classes ρ, n is the number of elements in the group, g_ρ is the number of operators in the ρth class, $\chi_{j\rho}$ is the character of the jth irreducible representation in the ρth class, and χ_ρ is the character of the physical quantity under consideration. This calculation of k_j is made now for S_2Cl_2.

$$k_1 = \frac{1}{2}\left[(1)(1)(3) + (1)(1)(-1)\right] = 1 \qquad (8.5)$$

$$k_2 = \frac{1}{2}\left[(1)(1)(3) + (1)(-1)(-1)\right] = 2 \qquad (8.6)$$

Thus $\chi_{m\rho}$ may be written as

$$\chi_{m\rho} = (1)\chi_{1p} + (2)\chi_{2p} \qquad (8.7)$$

A glance at the character table for S_2Cl_2 shows that this linear combination is correct for each of the two classes, and under a class χ_{mp} is indeed a linear combination of the irreducible representations in the class.

3 CALCULATION OF THE CHARACTERS OF THE ELECTRIC DIPOLE MOMENT FOR H_2O

The same calculations are done now for the molecule H_2O. The classes are $C_1 = E$, $C_2 = C_2$, $C_3 = \sigma_v$, $C_4 = \sigma_v$, and the character table is written above the dashed line.

	C_1	C_2	C_3	C_4
R_1	1	1	1	1
R_2	1	1	-1	-1
R_3	1	-1	1	-1
R_4	1	-1	-1	1
χ_{mp}	3	-1	1	1

Under each class the character of the dipole moment Eq. (8.3) is written under the dashed line where

$$\chi_{mp} = \sum_{j=1}^{4} k_j \, \chi_{jp} \tag{8.8}$$

The k_j's for H_2O are

$$k_1 = \frac{1}{4}\left[(1)(1)(3) + (1)(1)(-1) + (1)(1)(1) + (1)(1)(1)\right] = 1 \tag{8.9}$$

$$k_2 = \frac{1}{4}\left[(1)(1)(3) + (1)(1)(-1) + (1)(-1)(1) + (1)(-1)(1)\right] = 0 \tag{8.10}$$

$$k_3 = \frac{1}{4}\left[(1)(1)(3) + (1)(-1)(-1) + (1)(1)(1) + (1)(-1)(1)\right] = 1 \tag{8.11}$$

$$k_4 = \frac{1}{4}\left[(1)(1)(3) + (1)(-1)(-1) + (1)(-1)(1) + (1)(1)(1)\right] = 1. \tag{8.12}$$

Thus

$$\chi_{mp} = (1)\chi_{1p} + (0)\chi_{2p} + (1)\chi_{3p} + (1)\chi_{4p}. \tag{8.13}$$

This linear combination may be checked by substitution of the various characters for H_2O.

Because each component of the electric dipole moment of a molecule is a linear combination of displacements along that axis, the characters of the dipole moment under the covering operations will be those of the rectangular coordinates. The characters of the electric dipole moment may be summarized as follows:

$$\chi_{mE} = 3, \chi_{MC(\theta)} = 1 + 2\cos\theta, \chi_{MS(\theta)} = -1 + 2\cos\theta,$$

$$\chi_{MC_2(\theta)} = -1, \chi_{M\sigma_v(\theta)} = +1.$$

4 INFRARED SELECTION RULES

We shall use these results for H_2O and S_2Cl_2 to determine the infrared selection rules for these molecules, i.e., to learn which species of vibration are infrared allowed, but the rule stated below applies to *all* classes of symmetry. *In general, if a species (irreducible representation) appears in the reduced equation for χ_{mp}, then that species is infrared allowed.* Thus for S_2Cl_2, fundamental species A and B both appear in Eq.

(8.7) and therefore both species are infrared allowed. For H_2O, the reduced equations (8.13) for χ_{mp} does not contain $A_2(k_2=0)$ and so fundamental species A_1, B_1, and B_2 are infrared allowed and species A_2 is forbidden in the infrared.

The reasons for allowed and forbidden transitions may be understood from a quantum mechanical viewpoint. In quantum mechanics, the selection rule for a fundamental mode of vibration is determined by calculating the matrix element of the dipole moment. The value of the element is the integral $\int \Psi' M \Psi'' d\tau$. The integrand must be totally symmetrical (or of species A_1) and the mode is infrared-active if the integral does not vanish, i.e., the sign of the integral must be independent of the coordinate system in order for the integral not to vanish. The wave function Ψ' represents the initial vibrational state, Ψ'' represents the final vibrational state, and $d\tau$ is an element of volume in configuration space. The vibrational wave function is of the form

$$\Psi = N_v \, e^{-s^2/2} H_v(s) \tag{8.14}$$

where $H_v(s)$ is a Hermite polynomial which is either an even or an odd function for v even or odd, respectively. If the mode of vibration is non-degenerate, the effect of a covering operation on the coordinate "s" is multiplication by ±1, depending upon the character in the character table, so for v even or odd the wave function is multiplied by ±1. The lowest vibrational state is v = 0. In this state, the character of the vibrational wave function is +1 for all covering operations and the vibrational wave function belongs to species A_1. For a fundamental absorption transition, v = 0 → v = 1, Ψ' has v = 0 and Ψ'' has v = 1. The character of the integrad $\Psi' M_i \Psi''$ is the product of the individual characters, and the product $M_i \Psi''$ is ±1 under the covering operations. The sign of the integral must be independent of the coordinate system. Since the sign of Ψ' is always ±1, the sign of the product $M_i \Psi''$ must be + for every covering operation. This product can be + only if M_i and Ψ'' below to the same species (or if M and "s" belong to the same species).

The proof is longer for overtone or combination transitions, or for the Raman effect; but generally speaking, the reduction of the character of interest leads to the selection rules for infrared or Raman transitions.

5 CHARACTER OF THE POLARIZABILITY

The relationship between polarization and polarizability is expressed by

$$\bar{P} = \alpha \bar{E} \tag{8.15}$$

where $\bar{P}$ is the polarization (net electric dipole moment per unit volume), α is the polarizability, and $\bar{E}$ is the electric field. Depending upon the medium in which $\bar{P}$ exists, α may or may not be a constant.

If the nuclei and electrons in a molecule are placed in an electric field the electric charges are acted upon by a force which tends to displace the charges from their equilibrium positions, and the molecule acquires an induced electric dipole moment. If a coordinate system is fixed in the framework of the molecule then the components of the polarization are

$$P_x = \alpha_{xx}E_x + \alpha_{xy}E_y + \alpha_{xz}E_z \tag{8.16}$$

$$P_y = \alpha_{yx}E_x + \alpha_{yy}E_y + \alpha_{yz}E_z \tag{8.17}$$

$$P_z = \alpha_{zx}E_x + \alpha_{zy}E_y + \alpha_{zz}E_z \tag{8.18}$$

if the medium is *anisotropic*; that is, the electric field produces displacements along the field and at right angles to the field. It is necessary to find the characters of the polarizability under the various covering operations, as was done for the electric dipole moment, because the polarizabilities are important in determining which fundamental modes of vibration of the molecule will be Raman-active and how many modes of each type there will be. In order to find the characters of the polarizabilities α_{xx}, α_{xy},..., etc., first it is necessary to investigate the transformation properties of the subscripts of the α's, so we study the transformation of quadratic products of x, y, z, which are written in matrix form below.

$$A = \begin{bmatrix} x^2 & xy & xz \\ yx & y^2 & yz \\ zx & zy & z^2 \end{bmatrix} \tag{8.19}$$

Consider each of the covering operations and perform a unitary transformation by each operator, in turn, upon A. For the identity, $E\ A\ E^{-1} = A$. For $C(\theta)$,

$$B = C(\theta)\,A\,C(\theta)^+ = \begin{bmatrix} \cos\theta & \sin\theta & 0 \\ -\sin\theta & \cos\theta & 0 \\ 0 & 0 & 1 \end{bmatrix} \begin{bmatrix} x^2 & xy & xz \\ yx & y^2 & yz \\ zx & zy & z^2 \end{bmatrix} \begin{bmatrix} \cos\theta & -\sin\theta & 0 \\ \sin\theta & \cos\theta & 0 \\ 0 & 0 & 1 \end{bmatrix}$$

Let $a = \cos\theta$, $b = \sin\theta$, then

$$B = \begin{bmatrix} a^2x^2 + 2abxy + b^2y^2 & ,\ -abx^2 + a^2xy - b^2yx + aby^2 & ,\ azx + bzy \\ -abx^2 - b^2yx + a^2yx + aby^2 & ,\ b^2x^2 - abxy - abyx + ay^2 & ,\ ayz - bxz \\ azx + bzy & ,\ azy - bzx & ,\ z^2 \end{bmatrix} \tag{8.20}$$

Notice the symmetry in the matrix B: the xy-term is the same as the yx-term. These identical terms give the same information about the transformation, and only one term of each pair need be considered. The information contained in matrix B may be arranged in a tabular form: remember that the transformations are $x^2 \to x'^2$, $xy \to x'y'$, etc., and that the row and column indices in matrix B are those in matrix A except now we are talking about *primed* coordinates in Matrix B. So the terms x'^2, y'^2 and $x'y'$ may be written

$$\begin{bmatrix} x'^2 \\ y'^2 \\ x'y' \end{bmatrix} = \begin{bmatrix} a^2 & b^2 & 2ab \\ b^2 & a^2 & -2ab \\ -ab & ab & a^2-b^2 \end{bmatrix} \times \begin{bmatrix} x^2 \\ y^2 \\ xy \end{bmatrix} \tag{8.21}$$

and those terms involving $y'z'$ and $z'x'$ may be written as

$$\begin{bmatrix} y'z' \\ z'x' \end{bmatrix} = \begin{bmatrix} a & -b \\ b & a \end{bmatrix} \begin{bmatrix} yz \\ zx \end{bmatrix} \tag{8.22}$$

The remaining unitary transformations are

$$S(\theta) A\, S(\theta)^+$$
$$C_2(\theta) A\, C_2^+(\theta)$$
$$\sigma_v(\theta) A\, \sigma_v^+(\theta)$$

and their calculation is left as an exercise. All of this information is put into Table 8.2, which shows the relationships among the transformed quadratic products of x, y, z for each covering operation. This information yields the characters of the quadratic products and hence the polarizabilities which are also shown in the table, where $c = \cos 2\theta$, $d = \sin 2\theta$.

The individual characters of the polarizability under $C(\theta)$ and $S(\theta)$ are added to give a *general expression* for the character of the polarizability, $\chi_{\alpha p}$.

$$\chi_{\alpha p} = \overline{1} + 2\cos\theta + 4\cos^2\theta - 1$$
$$= \overline{2} + 2\cos\theta + 2\cos 2\theta \tag{8.23}$$

The minus sign goes with an improper rotation and the angle θ is zero for a reflection.

For a given molecule, the character of the polarizability under a class will be a linear combination of characters in the irreducible representations under that class, as are the characters of the electric dipole moment which was treated earlier.

Example—We proceed now to find the characters of the polarizability for S_2Cl_2 using Eq. (8.23) and the character table for this molecule:

	E	C_2
A	1	1
B	1	-1
$\chi_{\alpha p}$	6	2

and Eq. (8.23) becomes

$$\chi_{\alpha p} = k_1 \chi_{1p} + k_2 \chi_{2p}$$

Use Eq. (8.4) to find k_1 and k_2.

$$k_1 = \frac{1}{2}\left[(1)(1)(6) + (1)(1)(2)\right] = 4$$

$$k_2 = \frac{1}{2}\left[(1)(1)(6) + (1)(-1)(2)\right] = 2$$

Thus

Table 8.2 Transformation Matrices of Quadratic Products of x, y, z, Characters of Quadratic Products of x, y, z, and Characters of Polarizabilities.

Transformation Matrices

	E	$C(\theta)$	$S(\theta)$	$C_2(\theta)$	$\sigma_v(\theta)$
z^2	|1|	|1|	|1|	|1|	|1|
x^2 y^2 xy	$\begin{bmatrix} 1 & 0 & 0 \\ 0 & 1 & 0 \\ 0 & 0 & 1 \end{bmatrix}$	$\begin{bmatrix} a^2 & b^2 & 2ab \\ b^2 & a^2 & -2ab \\ -ab & ab & a^2-b^2 \end{bmatrix}$	$\begin{bmatrix} a^2 & b^2 & 2ab \\ b^2 & a^2 & -2ab \\ -ab & ab & a^2-b^2 \end{bmatrix}$	$\begin{bmatrix} c^2 & d^2 & 2cd \\ d^2 & c^2 & -2cd \\ cd & -cd & (d^2-c^2) \end{bmatrix}$	$\begin{bmatrix} c^2 & d^2 & 2cd \\ d^2 & c^2 & -2cd \\ cd & -cd & (d^2-c^2) \end{bmatrix}$
yz xz	$\begin{bmatrix} 1 & 0 \\ 0 & 1 \end{bmatrix}$	$\begin{bmatrix} a & -b \\ b & a \end{bmatrix}$	$\begin{bmatrix} -a & b \\ -b & -a \end{bmatrix}$	$\begin{bmatrix} c & -d \\ -d & -c \end{bmatrix}$	$\begin{bmatrix} -c & d \\ d & c \end{bmatrix}$

Characters of Quadratic Products

	E	$C(\theta)$	$S(\theta)$	$C_2(\theta)$	$\sigma_v(\theta)$
z^2	1	1	1	1	1
x^2, y^2, xy	3	$3a^2-b^2$	$3a^2-b^2$	c^2+d^2	c^2+d^2
yz, zx	2	$2a$	$-2a$	0	0

Characters of the Polarizabilities

	E	$C(\theta)$	$S(\theta)$	$C_2(\theta)$	$\sigma_v(\theta)$
α_{zz}	1	1	1	1	1
$\alpha_{xx}, \alpha_{yy}, \alpha_{xy}$	3	$4\cos^2\theta-1$	$4\cos^2\theta-1$	1	1
α_{yz}, α_{zx}	2	$2\cos\theta$	$-2\cos\theta$	0	0

$$\chi_{\alpha\rho} = 4\chi_{1\rho} + 2\chi_{2\rho} \tag{8.24}$$

6 RAMAN SELECTION RULES

The selection rules for fundamental modes of vibration which will appear in the Raman spectrum are obtained in a manner similar to that for the infrared selection rules. *If a species (irreducible representation) appears in the reduced equation for $\chi_{\alpha\rho}$, then that species is Raman allowed.* The fundamental species A and B both appear in Eq. (8.24) and therefore both species are Raman allowed.

Example—Calculate the Raman selection rules for H_2O.

	E	C_2	σ_v	σ_v
A_1	1	1	1	1
A_2	1	1	-1	-1
B_1	1	-1	1	-1
B_2	1	-1	-1	1
$\chi_{\alpha\rho}$	6	2	2	2

and Eq. (8.23) becomes

$$\chi_{\alpha\rho} = k_1\chi_{1\rho} + k_2\chi_{2\rho} + k_3\chi_{3\rho} + k_4\chi_{4\rho}$$

Use Eq. (8.4) to find the k's.

$$k_1 = \frac{1}{4}\left[(1)(1)(6) + (1)(1)(2) + (1)(1)(2) + (1)(1)(2)\right] = 3$$

$$k_2 = \frac{1}{4}\left[(1)(1)(6) + (1)(1)(2) + (1)(-1)(2) + (1)(-1)(2)\right] = 1$$

$$k_3 = \frac{1}{4}\left[(1)(1)(6) + (1)(-1)(2) + (1)(1)(2) + (1)(-1)(2)\right] = 1$$

$$k_4 = \frac{1}{4}\left[(1)(1)(6) + (1)(-1)(2) + (1)(-1)(2) + (1)(1)(2)\right] = 1$$

$$\chi_{\alpha\rho} = (3)\chi_\rho + (1)\chi_{2\rho} + (1)\chi_{3\rho} + (1)\chi_{4\rho} \tag{8.25}$$

Therefore fundamental modes of types A_1, A_2, B_1 and B_2 are Raman allowed.

7 CHARACTERS OF THE VIBRATIONAL COORDINATES

The group theoretic method of calculating infrared and Raman selection rules has been presented and now that we know how to calculate which vibrational species are infrared and Raman allowed, we learn here how to calculate the *number* of normal modes of vibration for each species. The number of normal modes of vibration for each species is obtained from characters of the vibrational coordinates of a molecule. We must know how many modes of vibration there are for each species, because the activity of the modes and hence the associated rotational fine structure depend upon the symmetry properties of the vibrations. We study molecular vibrations by regarding a molecule as a mass-spring system. The masses are chosen as the atomic nuclei because the electron masses are too small to make much of a contribution to the vibrational energy. The spring forces are the coupling forces between nuclei, and they arise from the electric fields of the nuclei and the electrons. A typical spring force for a molecule is of the order of a few millidynes per Angstrom. However, the exact relations between the masses and the field forces are seldom known, but it is assumed that the force field between the nuclei may be expanded in power series of small displacements of the nuclei from equilibrium. We study the behavior of these small displacement vectors under covering operations in order to obtain the *character of the displacement*, from which the number of normal modes of vibration for each species is obtained.

Consider an N-atomic molecule which undergoes a general motion in which the nth atom has displacement components of δx_n, δy_n, δz_n from equilibrium, relative to some space-fixed system of coordinates. When a covering operation is applied to the coordinate system, the displacement components of the nth atom are $\delta x'_n$, $\delta y'_n$, $\delta z'_n$, and these primed components will be linear combinations of the unprimed coordinates. For example, for the first atom

$$\delta x'_1 = a_{11}\delta x_1 + b_{11}\delta y_1 + c_{11}\delta z_1 + ...c_{1n}\delta z_n \tag{8.26}$$

$$\delta y_1 = a_{21}\delta x_1 + b_{21}\delta y_1 + ..., etc. \tag{8.27}$$

The character of this transformation is the sum of the diagonal elements of the matrix which describes the transformation. When a covering operation is used to transform the coordinate system, the operation exchanges the apparent position of equivalent atoms. Therefore, the displacement components of the nth atom may be written in terms of the mth atom only. The matrix of the transformation breaks up into sections, some of which have zero elements while other sections have non-zero elements. The atoms whose initial positions are not changed by the covering operation will yield diagonal elements in the matrix and thus contribute to the character of the displacement. Recall that for a displacement of one atom the character is

$$\chi_\rho = \pm 1 + 2 \cos \theta$$

If U_p and U_I are the number of atoms left unchanged by a proper and improper rotation, respectively, then the characters of the motion (vibrational coordinate) of the molecule are

$$\chi_\rho = U_p (1 + 2 \cos \theta), \text{ proper rotation} \tag{8.28}$$

$$\chi_I = U_I (-1 + 2 \cos \theta), \text{ improper rotation} . \tag{8.29}$$

Although the molecule is vibrating, its motion includes rotation and translation also, and these motions must be taken into account.

An N-atomic molecule has 3N degrees of freedom, not including spin. There are three degrees of translational freedom for non-linear or linear molecules, and there are three degrees of rotational freedom if the molecule is non-linear or two degrees of rotational freedom if the molecule is linear. Therefore the vibrational degrees of freedom for a non-linear molecule are 3N-6, whereas a linear molecule has 3N-5 degrees of freedom. The character of the vibrational motion only is obtained by subtraction of the characters of rotation and translation from the character of the general motion.

$$\chi_V = \chi_G - \chi_{rot.} - \chi_{trans.} \tag{8.30}$$

$$\chi_V = U_P(1 + 2\cos\theta) - (1 + 2\cos\theta) - (1 + 2\cos\theta)$$

$$= (U_p - 2)(1 + 2\cos\theta), \text{ proper rotation} \tag{8.31}$$

$$\chi_V = U_I(-1 + 2\cos\theta) - (1 - 2\cos\theta) - (-1 + 2\cos\theta)$$

$$= U_I(-1 + 2\cos\theta), \text{ improper rotation} \tag{8.32}$$

To find the number of normal modes of a species of vibration it is only necessary to substitute the proper values of the angle θ and reduce the character χ_{VP} or χ_{VI} by inspection or by the expansion theorem.

Example—Now we shall calculate the number of normal modes of each species of vibration for H_2O. The characters of the vibrational motion are calculated below the character table. Recall that σ_V is equivalent to an improper rotation by 0 degrees.

	E	C_2	σ_V	$\sigma_{V'}$
A_1	1	1	1	1
A_2	1	1	-1	-1
B_1	1	-1	1	-1
B_2	1	-1	-1	1
θ	0	π	0	0
$\cos\theta$	1	-1	1	1
$\pm 1 + 2\cos\theta$	3	-1	1	1
U_P	3	1	1	3
U_I	3	1	1	3
χ_{Vp}	3	1	1	3

The vibrational character under each class is reduced.

$$A_1 \text{ mode}: k_1 = \frac{1}{4}\Big[(1)(1)(3) + (1)(1)(1) + (1)(1)(1)\,(1)(1)(3)\Big] = 2$$

$$A_2 \text{ mode}: k_2 = \frac{1}{4}\Big[(1)(1)(3) + (1)(1)(1) + (1)(-1)(1) + (1)(-1)(3)\Big] = 0$$

$$B_1 \text{ mode}: k_3 = \frac{1}{4}\Big[(1)(1)(3) + (1)(-1)(3) + (1)(1)(3) + (1)(-1)(3)\Big] = 0$$

$$B_2 \text{ mode}: k_4 = \frac{1}{4}\Big[(1)(1)(3) + (1)(-1)(3) + (1)(-1)(3) + (1)(1)(3)\Big] = 1$$

Total = 3 fundamental modes

Thus there are two fundamental modes of vibration for species A_1 and one fundamental mode for species B_2 for a 3-atom molecule of symmetry C_{2v} or

$$\chi_{V\rho} = 2A_1 + B_2$$

as a check, $3N - 6 = 3(3) - 6 = 3$ fundamental modes of vibration. Recall that the character of the dipole moment for H_2O is reduced as

$$\chi_{M\rho} = A_1 + B_1 + B_2.$$

A comparison of the above two equations shows that species A_1 and B_2 are infrared-active, i.e., these modes will absorb infrared energy because $\chi_{V\rho}$ contains species which are components of the dipole moment character. In general, then, *a mode of vibration is infrared-active if it belongs to the same species as a component of the dipole moment*, i.e., we look for 1-1 correspondences between components of $\chi_{V\rho}$ and $\chi_{M\rho}$.

Notice that the species B_1 appears as a component of the dipole moment character and is indeed *allowed* for C_{2v} symmetry, but in this particular 3-atom molecule this mode is not *active* in the infrared.

Example—A different situation occurs if one considers another molecule of C_{2v} symmetry, CH_2F_2. In this case $N = 5$, $3N - 6 = 9$, and the reduction of the vibrational character yields

$A_1 \text{ mode}: k_1 = 4$

$A_2 \text{ mode}: k_2 = 1$

$B_1 \text{ mode}: k_3 = 2$

$B_2 \text{ mode}: k_4 = 2$

Total = 9 fundamental modes

The character of the dipole moment for C_{2v} symmetry does not contain A_2 as a component, so fundamentals of species A_1, B_1, and B_2 are infrared-active in CH_2F_2, and one expects to see infrared absorptions which correspond to $4A_1$ modes, $2B_2$ modes, and $2B_2$ modes.

Let us return to H_2O for another important concept. One may sketch intuitively what the three infrared-active modes should look like by looking at the character table for the characters of species A_1 and B_2. The A_1 mode is symmetric to all symmetry elements (all + 1's) and we know that there are two A_1 modes for H_2O. These A_1 modes are sketched below.

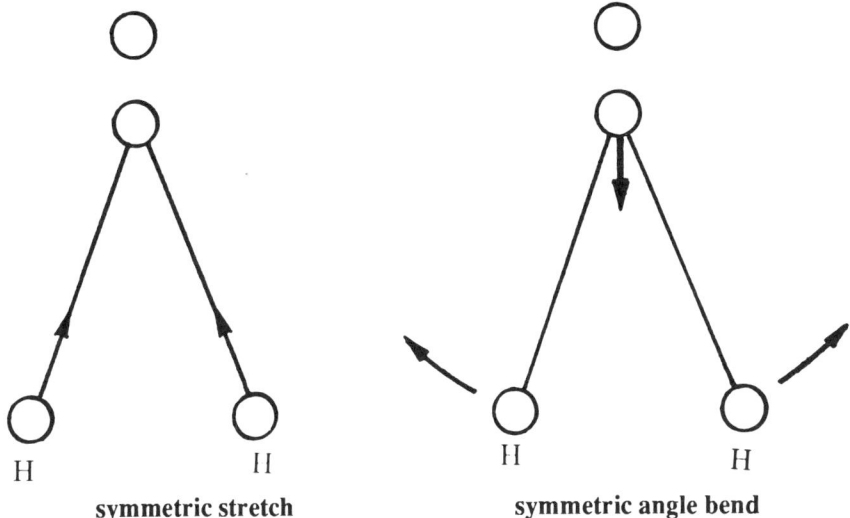

Figure 8.1 A_1 modes of H_2O.

Notice that these modes are symmetric with respect to E, C_2 σ_v, and $\sigma_{v'}$.

There is one B_2 mode, the antisymmetric stretch, and it is antisymmetric with

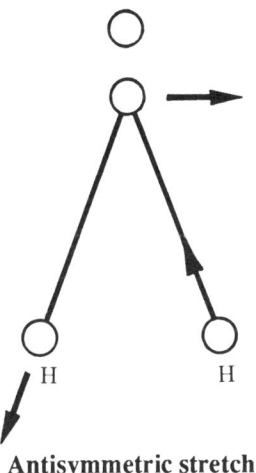

Figure 8.2 B_2 mode of H_2O.

respect to C_2 and σ_v (the -1's). In general, all modes must be constructed in such a manner that *momentum is conserved*.

We shall repeat these calculations for S_2Cl_2, $N = 4$ atoms.

	E	C_2
A	1	1
B	1	-1
θ	0	180°
$\cos\theta$	1	-1
$1+2\cos\theta$	3	-1
U_p	4	0
$\chi_{V\rho}$	6	2

The characters $\chi_{V\rho}$ are reduced:

$$k_1 = \frac{1}{2}\left[(1)(1)(6) + (1)(1)(2)\right] = 4$$

$$k_2 = \frac{1}{2}\left[(1)(1)(6) + (1)(-1)(2)\right] = 2$$

Total = 6 fundamental modes

The numbers k_1 and k_2 tell us that for S_2Cl_2 there are four fundamental modes of vibration of species A and two fundamental modes of vibration of species B. As a check, $3N - 6 = 3(4) - 6 = 6$ fundamental modes of vibration, which is what one expects for a 4-atom, nonlinear molecule. An earlier reduction of the characters of the dipole moment showed that species A and B are allowed in the infrared and Raman spectrum; therefore we expect to see six fundamental infrared absorptions and six fundamental Raman frequencies.

The fundamental modes of this molecule may be sketched as was done for H_2O by looking in the character table at the characters for species A and B. This sketch is left as an exercise for the student.

Infrared and Raman Selection Rules for Combination Transitions

A combination transition is the sum or difference of two fundamental frequencies, $v_i \pm v_j$. To obtain the selection rules for a combination it is necessary to multiply together the characters of the species to which v_i and v_j belong. For S_2Cl_2 suppose we want to learn if the combinations of vibration species A and B are infrared and Raman allowed; we must form the products of the characters of the species A and B to get the characters of the combination, $\chi_{C\rho}$. Consider all possible binary combinations; in this case there are three such possible combinations, $A \times A$, $A \times B$, and $B \times B$. The first and last combinations are the *second harmonics or first overtones* of species A and B; more will be said later about selection rules for overtones. The products of the characters are written below the character table for S_2Cl_2.

	E	C_2
A	1	1
B	1	-1
$\chi_{Cp}(A \times A)$	1	1
$\chi_{Cp}(A \times B)$	1	-1
$\chi_{Cp}(B \times B)$	1	1

Now proceed to reduce each of the characters by writing

$$\chi_{Cp}(A \times A) = k_1 \chi_{1p} + k_2 \chi_{2p}$$

$$\chi_{Cp}(A \times B) = k_3 \chi_{1p} + k_4 \chi_{2p}$$

$$\chi_{Cp}(B \times B) = k_5 \chi_{1p} + k_6 \chi_{2p}$$

where again the k's are calculated by Eq. (8.4), χ_{mp} being replaced by χ_{Cp}. However, for a simple character table like this one, inspection will suffice, and the results are:

(a) The characters of the combination $A \times A$ are the same as the characters for A ($A \times A = A$). Therefore, if frequency v_i, type A, combines with v_j, type A, the combination $v_i \pm v_j$ is infrared and Raman allowed *if A is allowed*.

(b) The characters of the combination $A \times B$ are the same as the characters for B ($A \times B = B$). Therefore if frequency v_i type B, combines with frequency v_j type B, the combinations $v_i \pm v_j$ is infrared and Raman allowed *if B is allowed*.

(c) The characters of the combination $B \times B$ are the same as the characters for A ($B \times B = A$). Therefore if frequency v_i type B, combines with v_j, type B, the combination $v_i \pm v_j$ is infrared and Raman allowed *if A is allowed*.

Table 8.3 - Infrared and Raman selection rules for combination modes of H_2O.

Combination	Combination Type	Raman	Infrared
$A_1 \times A_1$	A_1	allowed	allowed
$A_1 \times A_2$	A_2	allowed	forbidden
$A_1 \times B_1$	B_1	allowed	allowed
$A_1 \times B_2$	B_2	allowed	allowed
$A_2 \times A_2$	A_1	allowed	allowed
$A_2 \times B_1$	B_2	allowed	allowed
$A_2 \times B_2$	B_1	allowed	allowed
$B_1 \times B_1$	A_1	allowed	allowed
$B_1 \times B_2$	A_2	allowed	forbidden
$B_2 \times B_2$	A_1	allowed	allowed

These results are summarized below.

Combinations	A	B	Infrared	Raman
A×A	1	0	,,	,,
A×B	0	1	,,	,,
B×B	1	0	,,	,,

For H_2O:

The possible binary combinations of characters for H_2O are formed such as $A_1 \times A_1$, $A_1 \times A_2$, $A_1 \times B_1$, $A_1 \times B_2$, $A_2 \times A_2$, $A_2 \times B_1$, $A_2 \times B_2$, $B_1 \times B_1$, $B_1 \times B_2$, and $B_2 \times B_2$. The second unique combination, $A_1 \times A_2$ will be worked out; the remaining unique ones are left as an exercise, and the student may check his work by referring to Table 8.3 which summarizes the infrared and Raman selection rules for combinations in H_2O.

	E	C_2	σ_v	$\sigma_{v'}$
A_1	1	1	1	1
A_2	1	1	-1	-1
$\chi_{C\rho}(A_1 \times A_2)$	1	1	-1	-1

It is seen by inspection that the combination of two modes of types A_1 and type A_2 is mode type A_2. Note that for H_2O the fundamental mode type A_2 is Raman allowed only—therefore the combination mode $A_1 \times A_2$ is Raman allowed but not infrared allowed.

Selection Rules for Overtone Transitions

An overtone transition is analogous to a harmonic in radio frequency circuits. The infrared and Raman selection rules for the overtone (n-1) of a (nth harmonic) nondegenerate vibration are obtained by first raising to the nth power each character under the class of the fundamental mode type in question. If $\chi_{o\rho}$ is the character of the (n-1)th overtone under class ρ, then

$$\left[\chi_{j\rho}\right]^n = \chi_{o\rho}.$$

The $\chi_{o\rho}$ is reduced, as before, either by inspection or by using Eq. (8.4). To illustrate, we calculate the infrared and Raman selection rules for the 1st and 2nd overtones of S_2Cl_2, where $A^2 = A \times A$, $B^2 = B \times B$. The notation $\chi_{o\rho}(A^n)$ means the characters of the n-1 overtone of species A under class ρ.

USES OF THE CHARACTER TABLE 103

	E	C_2	
A	1	1	
B	1	-1	
$\chi_{op}(A^2)$	1	1	
$\chi_{op}(B^2)$	1	1	=> A type
$\chi_{op}(A^3)$	1	1	=> A type
$\chi_{op}(B^3)$	1	-1	=> B type

These calculations show that the character of the first overtone of a type A or B fundamental mode equals the character of the type A fundamental; these overtones are infrared and Raman allowed if the type A mode is infrared and Raman allowed. The A type fundamental mode in S_2Cl_2 is both infrared and Raman allowed, so both overtones are infrared and Raman allowed. By parallel reasoning, the second overtones of type A and B modes are infrared and Raman allowed. The selection rules for the overtones of S_2Cl_2 are summarized below.

Overtone	Character	Overtone Type	Raman	Infrared
n-1	$\chi_{op}(A^n)$	A	allowed	allowed
2n-1	$\chi_{op}(B^{2n})$	A	allowed	allowed
2n	$\chi_{op}(B^{2n+1})$	B	allowed	allowed

The selection rules for the 1st overtones of the type A_1 and A_2 fundamental modes of H_2O are calculated now.

	E	C_2	σ_v	$\sigma_{v'}$	
A_1	1	1	1	1	
A_2	1	1	-1	-1	
B_1	1	-1	1	-1	
B_2	1	-1	-1	1	
$\chi_{op}(A_1^2)$	1	1	1	1	$\to A_1$ type
$\chi_{op}(A_2^2)$	1	1	1	1	$\to A_1$ type

In each case, the character of the first overtone of fundamental modes A_1 and A_2 is of A_1 types, so each of these overtones is infrared and Raman allowed in H_2O, as seen earlier. The selection rules for the overtones of H_2O are summarized below.

n	overtone	overtone type	Raman	Infrared
even and odd	A_1^n	A_1	allowed	allowed
even	A_2^n	A_1	allowed	allowed
odd	A_2^n	A_2	allowed	forbidden
even	B_1^n	A_1	allowed	allowed
odd	B_1^n	B_1	allowed	allowed
even	B_2^n	A_1	allowed	allowed
odd	B_2^n	B_2	allowed	allowed

It is more involved to calculate the overtone selection rules for doubly degenerate or triply degenerate vibrations. The degeneracy is known from the character table by inspection of the first class which contains only the identity. If the character "2" appears under the identity class, then the species which contains that "2" as one of its characters is a doubly degenerate species. A similar statement can be made for triple degeneracy. Thus for doubly degenerate vibrations (type E), the character of the overtones $\chi_{E\rho}^n$, is given by

$$\chi_{E\rho}^n = \frac{1}{2}\left[\chi_{E\rho}^{n-1}\chi_{E\rho} + \chi_{E\rho^2}\right]$$

where $\chi_{E\rho}$ is the character of the type E vibration under class ρ, and $\chi_{E\rho^2}$ is the character corresponding to a sample operation under class ρ, performed *twice* in succession. As an example, the character table for C_{3v} symmetry is

	C_1 (E)	C_2 (C_3)	C_3 (σ_v)
A_1	1	1	1
A_2	1	1	-1
E	2	-1	0

and the identity operation performed twice in succession leaves a configuration which is indistinguishable from having performed it once, so $\chi_{E1} = \chi_{E1^2} = 2$. For c_2, $\theta = 120°$, so to perform his operation twice is equivalent to rotating the molecule about the symmetry axis by $+120°$ and $\chi_{E2^2} = \chi_{E2} = -1$.

$$k_2 = \frac{1}{6}\left[(1)(1)(3) + (2)(1)(0) + (3)(-1)(1)\right] = 0$$

$$k_3 = \frac{1}{6}\left[(1)(2)(3) + (2)(-1)(0) + (3)(0)(1)\right] = 1$$

Thus

$$\chi^2_{E\rho} = \chi_{1\rho} + \chi_{3\rho}$$

or $\quad E^2 = A_1 + E$

This result is checked by adding the characters of species A_1 and E.

The selection rule for this overtone is:

For C_{3v} symmetry, the first overtone of a type E vibration is infrared and Raman allowed if fundamental modes A_1 and E are infrared and Raman allowed. The selection rules for fundamental modes of C_{3v} symmetry do allow A_1 and E type modes in the infrared and Raman.

The overtones of triply degenerate frequencies (type F) are calculated from the equation

$$\chi^n_{F\rho} = \frac{1}{3}\left\{2\chi_{F\rho}\chi^{n-1}_{F\rho} - \frac{1}{2}\chi^{n-2}_{F\rho}\left[\chi_{F\rho}\right]^2 + \frac{1}{2}\chi_{F\rho_2}\chi^{n-2}_{F\rho} + \chi_{F\rho^*}\right\}$$

where $\chi^n_{F\rho}$ is the character of the (n-1)st overtone, $\chi_{F\rho}$ is the character of species F under class ρ, $\chi^{n-1}_{F\rho}$ and $\chi^{n-2}_{F\rho}$ are the characters of the (n-2) and (n-3) overtone, respectively, and $\chi_{F\rho^*}$ is the character corresponding to a sample operation under class ρ performed n times in succession. For higher degeneracies, see the paper by Tisa.[1] For c_3, performing a reflection twice in a vertical plane of symmetry is the same as the identity operation, so $\chi_{E3^2} = \chi_{E1} = 2$. Therefore, under the character table we write the calculation of $\chi^2_{E\rho}$. This character is reduced in the usual fashion to obtain the selection rules, i.e.,

	C_1 (E)	C_2 (C_3)	C_3 (σ_v)
A_1	1	1	1
A_2	1	1	-1
E	2	-1	0
$(\chi_{E\rho})^2$	4	1	0
$\chi_{E\rho^2}$	2	-1	2
$\chi^2_{E\rho}$	3	0	1

Consider only this row to calculate → $\chi^n_{E\rho}$

$$\chi^2_{E\rho} = k_1\chi_{1\rho} + k_2\chi_{2\rho} + k_3\chi_{3\rho}$$

where

$$k_j = \frac{1}{n}\sum_\rho g_\rho \chi_{j\rho}\chi^2_{E\rho}$$

There are six operations in this group (E; two possible rotations by $\pm\frac{2\pi}{3}$; three equivalent vertical reflection planes) so in the reduction theorem, $n = 6$, class C_1

contains 1 operator, class C_2 contains two operators, and class C_3 contains three operators, i.e., $g_1 = 1, g_2 = 2, g_3 = 3$.

$$k_1 = \frac{1}{6}\left[(1)(1)(3) + (2)(1)(0) + (3)(1)(1)\right] = 1$$

$$k_2 = \frac{1}{6}\left[(1)(1)(3) + (2)(1)(0) + (3)(-1)(1)\right] = 0$$

$$k_3 = \frac{1}{6}\left[(1)(2)(3) + (2)(-1)(0) + (3)(0)(1)\right] = 1$$

Thus

$$\chi^2_{E\rho} = \chi_{1\rho} + \chi_{3\rho}$$

or $\uparrow \quad \uparrow \quad \uparrow$

$$E^2 = A_1 + E$$

This result is checked by adding the characters of species A_1 and E.

The selection rule for this overtone is:

For C_{3v} symmetry, the first overtone of a type E vibration is infrared and Raman allowed if fundamental modes A_1 and E are infrared and Raman allowed. (The selection rules for fundamental modes of C_{3v} symmetry allow A_1 and E type modes in the infrared and Raman.)

The overtones of triply degenerate frequencies (type F) are calculated from the equation

$$\chi^n_{F\rho} = \frac{1}{3}\left\{2\chi_{F\rho}\chi^{n-1}_{F\rho} - \frac{1}{2}\chi^{n-2}_{F\rho}\left[\chi_{F\rho}\right]^2 + \frac{1}{2}\chi_{F\rho^2}\chi^{n-2}_{F\rho} + \chi_{F\rho^n}\right\}$$

where $\chi^n_{F\rho}$ is the character of the (n-1)st overtone, $\chi_{F\rho}$ is the character of species F underclass ρ, $\chi^{n-1}_{F\rho}$ and $\chi^{n-2}_{F\rho}$ are the characters of the (n-2)nd and (n-3)rd overtone, respectively, and $\chi_{F\rho^n}$ is the character corresponding to a sample operation under class ρ performed n times in succession. For higher degenericies, see the paper by Tisa.[1]

The selection rules for overtones and combinations may be misleading for the following reasons unless caution is observed.

Overtones

In nonlinear electrical circuits (which are often used as harmonic generators) one expects to see harmonics at integral multiples of the fundamental. Molecular harmonics which are infrared active usually do not fall at an integral multiple of the fundamental mode because of anharmonicity in the potential energy function of the molecule. The discrepancy is not large, but it cannot be ignored. Like the circuit case,

---

[1] L. Tisa, Ziets, F. Physik, **82**, 48 (1933).

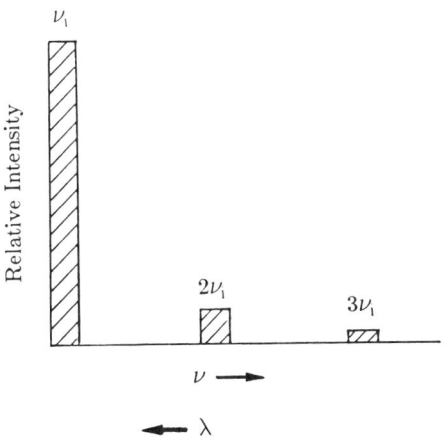

Figure 8.3 Relative intensities of fundamental, first overtone and second overtone.

the intensity of the molecular overtone is weaker than the intensity of the fundamental. In the molecular case, the overtones are very weak in intensity when compared to the fundamental intensity. An approximate idea of relative intensity may be seen from Fig. 8.3. Also it may happen that a fundamental is not infrared active but one or more of its overtones are infrared active.

When a band appears in a Raman spectrum, however, most often it is a fundamental mode; but there are cases which yield observable overtone spectra. It is possible also that a band may be active but not be seen in a spectrum, simply because its intensity is too weak to be recorded above the noise.

Combinations

The combination $v_i + v_j$ does not usually occur at the arithmetic sum of the fundamental modes v_i and v_j because of anharmonicity. The discrepancy is not large, but it is significant. Combinations of overtones and fundamentals may occur also, as may tertiary combinations such as $v_1 + v_2 + v_3$: their selection rules are calculated in analogous fashion.[2] An example is D_2O vapor in which $v_1 + v_2 + v_3 = 6538\ cm^{-1}$ and HDO vapor where $2v_1 + v_3 = 9050\ cm^{-1}$.

If active, the combinations $v_i - v_j$ will agree with the frequency observed in the recorded infrared spectrum because anharmonics effect cancel.

REFERENCES

1. L. Tisa, Ziets. F. Physik **82**, 48 (1933).
2. G. Herzberg, *Molecular Spectra and Molecular Structure II. Infrared and Raman Spectra of Polyatomic Molecules*, D. Van Nostrand Co., Inc. N.Y. (1945).

CHAPTER

9

LASER FUNDAMENTALS

The word laser is an acronym for *L*ight *A*mplification by the *S*timulated *E*mission of *R*adiation. The word "light" also refers to wavelengths that are invisible such as infrared radiation emitted by a molecular laser. It is these lasers that are discussed in this chapter.

All lasers are composed of: (a) a medium capable of lasing (b) an optical resonant cavity, and (c) a "pump" which provides excitation to the medium in order to achieve population inversion between a chosen pair of energy levels. All "molecular" lasers are gas lasers but not all gas lasers emit infrared radiation. Helium-neon and argon ion lasers are gas lasers but they have emissions in the visible portion of the spectrum also. We will discuss (a) and (b) above in some detail, with less emphasis on (c). The pumping of a gas laser is usually achieved by creating an electrical discharge in the gas medium.

Many different gases can be made to lase. From an engineering point of view intelligent choices of gases are made from detailed knowledge about the molecular energy states of a particular gas. This knowledge must include energy spacings (ΔE) and state lifetimes. No mention of state lifetimes has been made but it is a parameter obtained by observing spectroscopically the rate of decay from a higher energy state to a lower energy state. Relatively long ($\approx 10^{-8} sec$) upper state lifetimes are desirable for lasing action.

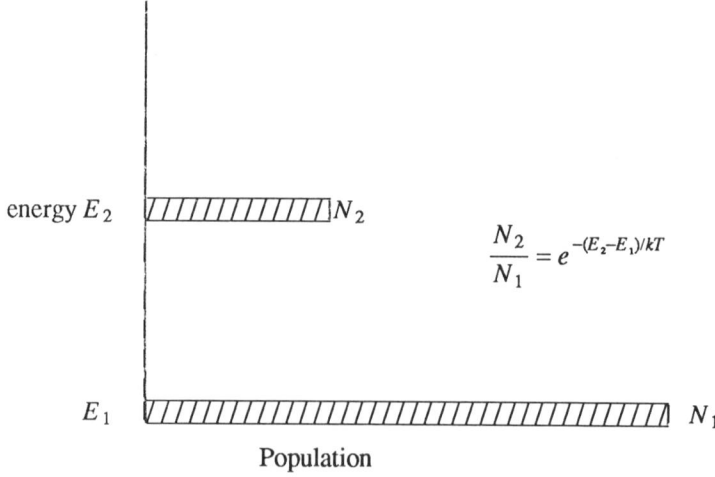

Figure 9.1 A two-level system in thermodynamic equilibrium.

Amplification of electromagnetic energy is not possible using a gaseous molecular medium in thermodynamic equilibrium with its surroundings because the Boltzmann distribution of the state populations requires there be more molecules in the lower states than in the upper states. (These states for a molecular gas laser are the various energy levels that define the allowed vibration, vibration-rotation, and rotation energy levels). As a result, if electromagnetic energy is passed through the gas the net result is always absorption of energy and no amplification is possible.

Consider two energy levels chosen from a multi-level system. A simple two-level system is shown in Fig. 9.1 and based upon the Boltzmann factor $e^{-h\nu/kT}$, the upper level has a smaller number of molecules in it than the lower level. If this system is in thermal equilibrium ($N_2 < N_1$), the number of upward transitions (absorption) must balance the number of downward transitions (emission) so no photon amplification occurs in this system.

For lasing to occur, equilibrium must be destroyed by pumping molecules from N_1 to N_2 and a population inversion established so that $N_2 > N_1$. Even if this condition is achieved the medium will constantly tend to return to equilibrium ($N_2 < N_1$) and a pump must be used continuously to maintain the inversion.

For $N_2 > N_1$ *linear, phase coherent* amplification of a photon becomes possible but amplification ceases when $N_2 = N_1$. This process, called *stimulated emission*, has a wave vector $\underline{k}$ and a polarization direction which are the same as those of the photons already present. Spontaneous emission is always present; it is independent of the external conditions and will occur even if no radiation is allowed to reach the molecular system.

Not all energy levels exhibit the same response to an external signal nor is the rate of energy decay to a lower level the same for all levels. Levels that exhibit relatively long state lifetimes are good candidates for lasing. This condition is characterized by a number, the oscillator strength of the level, which is related reciprocally to

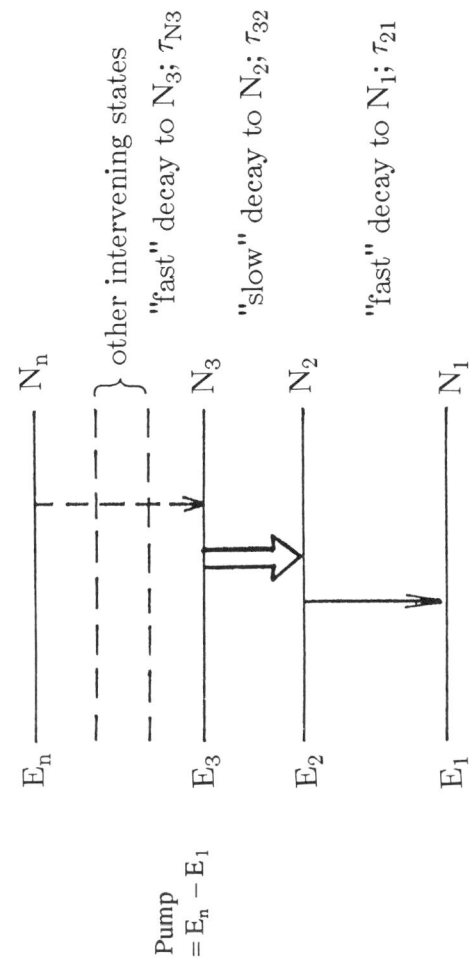

Figure 9.2 An N-level system pumped by a laser.

the radiative decay time of that level. Levels with weak oscillator strengths are desirable because the rate of population change from this level by spontaneous emission is less than from a level whose rate of energy decay is faster (stronger oscillator strength). The pump power required to maintain the population difference between two levels is less for the weaker oscillator strength. Population inversion between two states cannot be achieved by resonant absorption of photons corresponding to the transition frequency between the two energy states. However, inversion can be achieved by pumping to some level higher in energy than N_2, then with knowledge of the state lifetimes τ, choose which pairs of states can achieve a relative population inversion. The process is indicated in Fig. 9.2 for an N-level system. The population N_3 becomes greater than N_2 and population inversion is achieved. The process is described by rate equations in $\dfrac{dN_n}{dt}$ which can be solved to yield the population N_3 relative to N_2.

The radiation from an inverted population of molecules in the presence of an external radiation field consists of *two* parts, spontaneous radiation and stimulated radiation. The intensity of the spontaneous radiation is proportional only to the number of molecules in the upper state and it is both phase and direction *independent* of the external radiation field, whereas the phase and direction of the stimulated radiation is the same as that of the stimulating external radiation. Also, the probability of the stimulated transition occurring is proportional to the energy density of the external field.

The reader who wants to explore these concepts more fully is encouraged to study about (a) Einstein's A and B coefficients which are related to spontaneous and stimulated emission probabilities and (b) the quantum theory of light. The fundamental differences between spontaneous and stimulated emission are based in quantum theory, and the theory treats the electric susceptibility $\chi_e(\omega)$ in media (such as gaseous molecules) that have many resonances. The mechanism that gives rise to χ_e is the interaction of an electric field and the electric dipole moment, $\underline{E} \cdot \underline{p}$. The frequency of the field may correspond to visible light or to infrared, but the energy transfer from field to atom (or molecule) occurs through this interaction. This energy is transferred via electrons to vibrational and/or rotational motion of the molecule. Recall from Chapter 2 that the quantum state of a molecule at any instant is described by its wave function which is a superposition of the molecule's characteristic energy levels in proportions given by the wave function amplitude coefficient $c_n(t)$, i.e.,

$$\psi(t) = \sum_n c_n(t)\psi_n . \qquad (9.1)$$

The coefficients $c_n(t)$ are complex numbers and they are time functions whether relaxation processes (spontaneous) or external signals are present, but the ψ_n are not time dependent. The numbers $|c_n|^2$ yield the probability that a measurement made upon the collection of molecules will give the energy E_n. Our ideas of discrete "jumps" between two energy levels m and n is really only qualitative. The quantum theory teaches that when an external signal is applied to a molecule (or atom) the wave function $\psi(t)$ evolves in time continuously according to Schrodinger's time-dependent equation. If the frequency of the external signal is at or near the frequency f_{mn} of a transition between the energy levels E_m and E_n, then the coefficients $c_m(t)$ and $c_n(t)$ will change more rapidly than the other coefficients. Therefore the probabilities $|c_m(t)|^2$ and $|c_n(t)|^2$ will also change more rapidly compared to the other probabilities. This means that in a large collection of molecules (or atoms) the populations N_m and N_n will therefore change. It is *as if* some of the molecules in the collection "jumped" from level m to level n, but actually all the states of the molecules evolve in time, some more than others.

A classical description of this lasing process gives insight into the quantum situation. Consider an oscillating ensemble of charges coupled through the electric dipole moment to an external, time-varying elective field $E(t)$. This process can be described though the relation between the electric dipole moment/unit volume and the electric field

$$\underline{P} = Nxe = \varepsilon_o \chi_e \underline{E} \tag{9.2}$$

where $\underline{P}$ is the polarization vector, ε_o is the dielectric constant of free space and χ_e is the electric susceptibility. Refer to Eq. (4.44) reproduced here.

$$\chi(\omega) = \frac{jeE_x(\omega)}{u \ell \omega} \cdot \frac{1}{1 + 2j\frac{(\omega - \omega_0)}{\ell}} \tag{9.3}$$

This equation expresses the mass displacement *vs* frequency of an oscillator modeled as a mass-spring system. A molecule can be modeled classically in this way, but for more than one mass and spring, we expect more than one resonant frequency. Consider a molecule such as HCl (which can be made to lase). Set its reduced mass be m and let the rate of energy decay γ for HCl replace the quantity ℓ in Eq. (4.44).

The equation is now

$$\chi(\omega) = \frac{jeE_x(\omega)}{m\gamma\omega} \cdot \frac{1}{1 + 2j\frac{(\omega - \omega_o)}{\gamma}} \tag{9.4}$$

and in the frequency domain

$$P(\omega) = Nx(\omega)e = \varepsilon_o \chi_e(\omega) E(\omega). \tag{9.5}$$

Substitute for $X(\omega)$ and solve for $\chi_e(\omega)$

$$\chi_e(\omega) = \frac{je^2 N_v}{m\gamma\omega\varepsilon_o} \cdot \frac{1}{1 + 2j\frac{(\omega - \omega_c)}{\gamma}}. \tag{9.6}$$

For molecules having more complicated geometry, there is a sum of terms for $\chi_e(\omega)$, each with a reduced mass belonging to that part of the molecule executing a mode under consideration.

Although this equation describes an electric field interacting with a vibrating mass-charge system, one can just as easily consider the interaction of the electric field with a *vibrating-rotating* electric dipole and get analogous results, then by superposition add the vibration and rotation terms to get the response of the molecule to a polychomatic electric field. An applied electric field at frequency ω would then determine which of the term values are large enough to consider in a calculation of $\chi_e(\omega)$.

A quantum mechanical treatment of the interaction of a polychomatic electric field and a molecule is a calculation requiring time-dependent perturbation theory but these results have many similarities to the classical result. It turns out that the electric electric susceptibility is still written as a function of frequency but it is described for molecular transitions between two quantum states, m and n (m>n), assuming that the transition $m \to n$ is allowed by the selection rules, viz.,

$$\chi_e(\omega)_{mn} = \frac{j3F_{mn}(N_n - N_m)e^2}{m\varepsilon_o \omega_{mn} \Delta\omega_{mn}} \cdot \frac{1}{1+2j\frac{(\omega - \omega_{mn})}{\Delta\omega_{mn}}} \tag{9.7}$$

In this equation:
F_{mn} is the oscillator strength of transition $m \to n$.
N_n is the population of the lower level.
N_m is the population of the upper level.
ω_{mn} is the transition frequency, $\omega_{mn} = \dfrac{(E_m - E_n)}{h}$
$\Delta\omega_{mn}$ is the linewidth (as defined in Chapter 3) of the transition $m \to n$.
This function, when separated into real and imaginary parts becomes

$$\chi_{mn}(\omega) = \chi'_{mn}(\omega) + j\chi''_{mn}(\omega) \tag{9.8}$$

$$\chi'_{mn}(\omega) = \frac{2\left[\dfrac{3F_{mn}(N_n - N_m)e^2}{m\varepsilon_o \omega_{mn} \Delta\omega_{mn}}\right]e^2 \left[\dfrac{\omega - \omega_{mn}}{\Delta\omega_{mn}}\right]}{1 + \left[\dfrac{2(\omega - \omega_{mn})}{\Delta\omega_{mn}}\right]^2} \tag{9.9}$$

$$\chi''_{mn}(\omega) = \frac{\left[\dfrac{3F_{mn}(N_n - N_m)e^2}{m\varepsilon_o \omega_{mn} \Delta\omega_{mn}}\right]}{1 + \left[\dfrac{2(\omega - \omega_{mn})}{\Delta\omega_{mn}}\right]^2} \tag{9.10}$$

Notice that (a) for $N_n - N_m > 0$ (thermal equilibrium) $\chi''_{mn}(\omega)$ is positive (absorption of energy) for all ω and $\chi'_{mn}(\omega)$ is negative for $\omega < \omega_{mn}$ (see Fig. 9.3); (b) if $N_n - N_m < 0$ (population inversion) then the signs of $\chi''_{mn}(\omega)$ and $\chi'_{mn}(\omega)$ are reversed and $\chi''_{mn}(\omega)$ represents emission along with a 180° phase change in $\chi'_{mn}(\omega)$.

An analogy can be made with an electrical circuit having a complex admittance that describes an RLC parallel configuration. If the capacitor has its dielectric $\varepsilon = \varepsilon_o(1 + \chi_{emn})$ modified to include $\chi_{mn}(\omega)$, then there will be a pulling of the frequency of the original circuit because of a change in the capacitance value. A "loaded" cavity containing gas molecules behaves similarly in that the change with

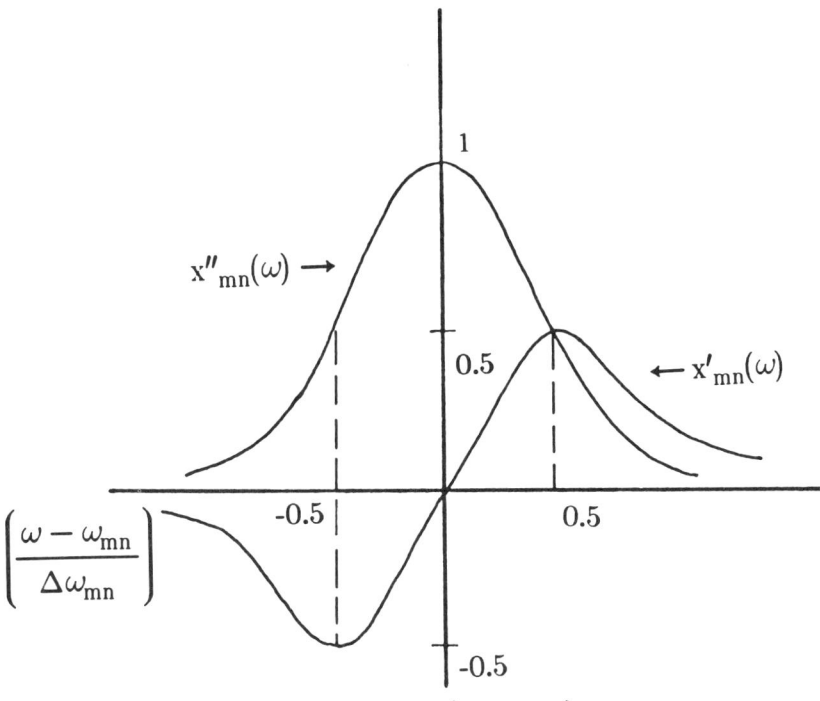

Figure 9.3 Real and imaginary parts of χ_{mn} vs $\left[\dfrac{\omega - \omega_{mn}}{\Delta\omega_{mn}}\right]$

frequency of $\chi_{mn}(\omega)$ will pull the resonant cavity frequency. A gas having population inversion between two energy levels describing (a) vibration, or (b) rotation, or (c) vibration-rotation can be made to provide gain (linear, phase-coherent amplification of photons). The gas must be placed into a resonant cavity that provides (a) low-loss optical feedback, and (b) a pump to create and maintain population inversion. The amplification process is linear because each photon created may trigger another photon: in a resonant cavity, the process is cascaded by mirrors at each end of the cavity, thus providing many passes of the photons thru the gas. The optical system is a closed loop and, like its electronic counterpart, oscillation may occur when the open loop gain is at least unity and the phase shift around the loop (one round trip) is an integral multiple of 2π. Because the optical cavity is many wavelengths long, the frequencies that occur within the linewidth of a lasing molecular transition can cause the roundtrip cavity phase shift to be large integral multiples of 2π. For example, the "optical path" in the cavity is defined to be $2nL$ where n is the refractive index of the gas and $2L$ is the distance for one round trip. The quantity $(2nL)\beta$ is the phase shift, Φ, for the path L.

$$2nL\,\beta = \Phi(= 2\pi m) \tag{9.11}$$

Write the phase constant β as $\dfrac{2\pi}{\lambda}$, then

$$2nL\beta = 2nL\left[\frac{2\pi}{\lambda}\right] = 2\pi\, m$$

or

$$\frac{2nL}{\lambda} = m\ . \tag{9.12}$$

The order of magnitude of m can be estimated as follows: Let L = 1 meter, n = 1, λ = 10.6 µm, then m is of the order of 10^5. Thus there are many half wavelengths in the standing wave between the end mirrors of an axial cavity. These wavelengths are called the axial modes of the cavity and gain will occur each time m increases by one. The frequency spacing between the two adjacent modes can be calculated from

$$f_{m+1} - f_m = \frac{c}{\lambda_{m+1}} - \frac{c}{\lambda_m} \tag{9.13}$$

$$= \frac{c(m+1)}{2nL} - \frac{c(m)}{2nL} = \frac{c}{2nL}(m+1-m)$$

$$= \frac{c}{2nL}\ .$$

If $L = 1$ meter and $n = 1$, then $f_{m+1} - f_m = 150\,MHz$. Thus there can be many sharply peaked resonances within a molecular linewidth. Other cavity lengths and refractive indices will vary this difference somewhat, but it remains on the order of 10^6 Hz.

A mode frequency can be selected and others rejected by placing a grating in the cavity so that the desired frequency is reflected back to produce gain. The frequency reflected from the grating is a function of the groove spacing of the grating, d, the angle of incidence, θ, and the incident wavelength, λ, according to

$$n\lambda = 2\,d\sin\theta\ ,\ n\ is\ an\ integer\ called\ the\ order\ .$$

CHAPTER
10

MOLECULAR LASERS

1 THE CO_2 LASER

The CO_2 laser is a good example of lasing on vibration-rotation transitions. Spectroscopic knowledge is put to practical use in the design of molecular lasers.

The possibility of using molecular vibration transitions for infrared lasers was discussed as early as 1961, and since that time many molecular systems have been made to lase. Perhaps the best known molecular laser is the carbon dioxide laser and its most useful output wavelength is 10.6 μm. Such radiation is useful in heating and cutting applications. In optical communications systems the 10.6 μm radiation propagates readily through earth's atmosphere and thus is useful in space communications.

There are many types of gas lasers available today and each has unique wavelength(s). A choice of lasers is based upon the engineering application. One of the most versatile gas lasers is the carbon dioxide laser. Its versatility is based upon an uncommon blend of attributes that make it useful in cutting, welding, communications, heat treating, etc. Cost and ease of use are important considerations for a prospective user; the CO_2 laser is relatively low cost and it is easy to maintain and operate, even at high power levels.

Typical industrial CO_2 lasers range in power from 500 W to 2 kW and the input powers required to operate them vary from 1 kVA to 30 kVA. These lasers have high efficiency and good beam quality. They can be sharply focused in extremely small areas for cutting and welding in microcircuits.

The desirable features of the laser stem from the low energy (0.2-0.3 eV) of the molecular states used to produce lasing which, in turn, reduces spontaneous emission noise. Also, the pump for this laser may be dc, ac, or radio frequency energy and the laser may be operated CW or pulse.

Carbon dioxide is a linear molecule with 3N-5 normal modes of vibration. The carbon atom is situated between the two oxygen atoms. This linear structure has three fundamental modes of vibration—the symmetric stretch, the antisymmetric stretch, and the angle-bending mode which is doubly degenerate. These modes are illustrated in Fig. 10.1.

The radiation from a CO_2 laser is composed of photons emitted when the molecule's vibrational energy decays from the asymmetric stretch vibrational level to the symmetric stretch vibrational level as depicted in Fig. 10.1. Each vibrational mode is quantized and the vibrational state is designated by three integers shown in parentheses in Fig. 10.1. The first integer is the vibrational quantum number of the symmetric mode, the second integer is the vibrational quantum number of the bending mode, and the third integer is the vibrational quantum number of the antisymmetric mode. Modes v_3 and v_2 are infrared absorbing because these molecular motions produce a change in the electric dipole moment of the molecule, whereas mode v_1 does not absorb infrared energy because the symmetry of the mode produces no change in the electric dipole moment.

The relative energies of these modes are shown in Fig. 10.2 but mixed vibrational states do exist as well, for example (111). The actual maximum efficiency realized from this lasing transition is about 20% which is good for any laser. This efficiency is a measure of the power input to the gas converted to output beam power.

There are rotational energy levels sandwiched between the two vibrational levels used for lasing, so *vibration-rotation* transitions are the actual transitions that produce laser photons. Only a select few of the vibration-rotation transitions lase for reasons discussed earlier in Chapter 9. The P(18) transition is the most powerful.

De-excitation of the lower vibrational level occurs through normal decay processes. However, if the CO_2 gas becomes too hot (>150°C) the lower levels become populated again (Boltzmann factor) so the waste heat must be removed by conduction (in order to depopulate these levels), by continuing replacement of the gas with cool gas, or by a cooling jacket for the gas discharge tube. Depending upon the laser power level, the gas may be moved at high pressure and velocity, or at moderate velocity, or at a trickle. Conduction cooled CO_2 lasers are available up to CW powers of 1 kW and flow cooled CO_2 lasers range up to 15 kW CW power. The flow cooled CO_2 laser can be made shorter and it is a more compact laser than the conduction cooled variety.

Pulses can be produced at peak power several times higher than the continuous wave capability of the laser. The technique is called superpulsing. The superpulses

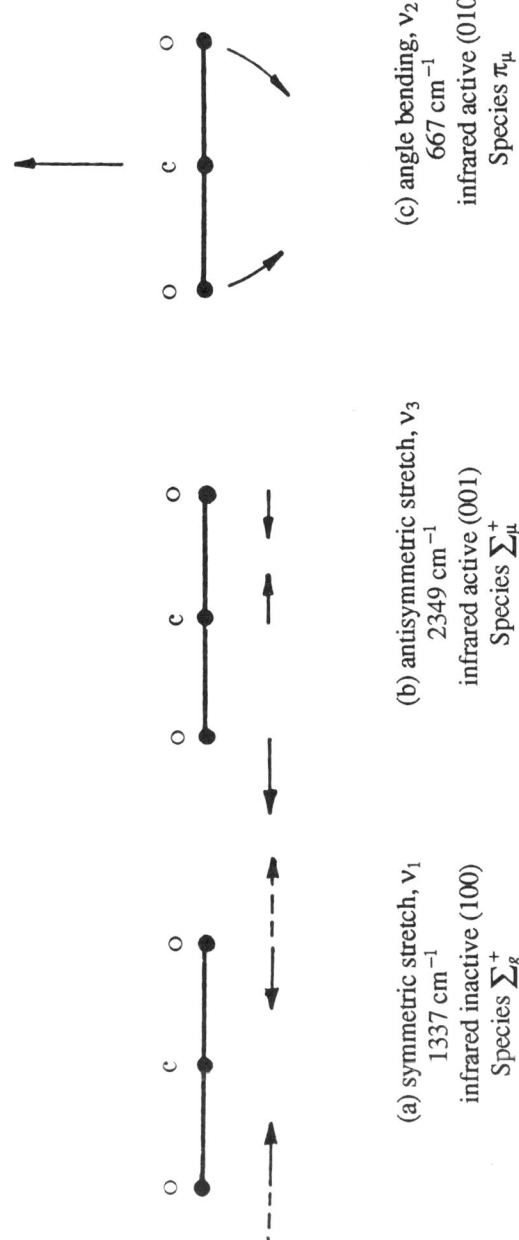

Figure 10.1 Fundamental vibrational modes of CO_2. 1 cm^{-1} = 30 GHz.

(a) symmetric stretch, ν_1
1337 cm^{-1}
infrared inactive (100)
Species Σ_g^+

(b) antisymmetric stretch, ν_3
2349 cm^{-1}
infrared active (001)
Species Σ_μ^+

(c) angle bending, ν_2
667 cm^{-1}
infrared active (010)
Species π_μ

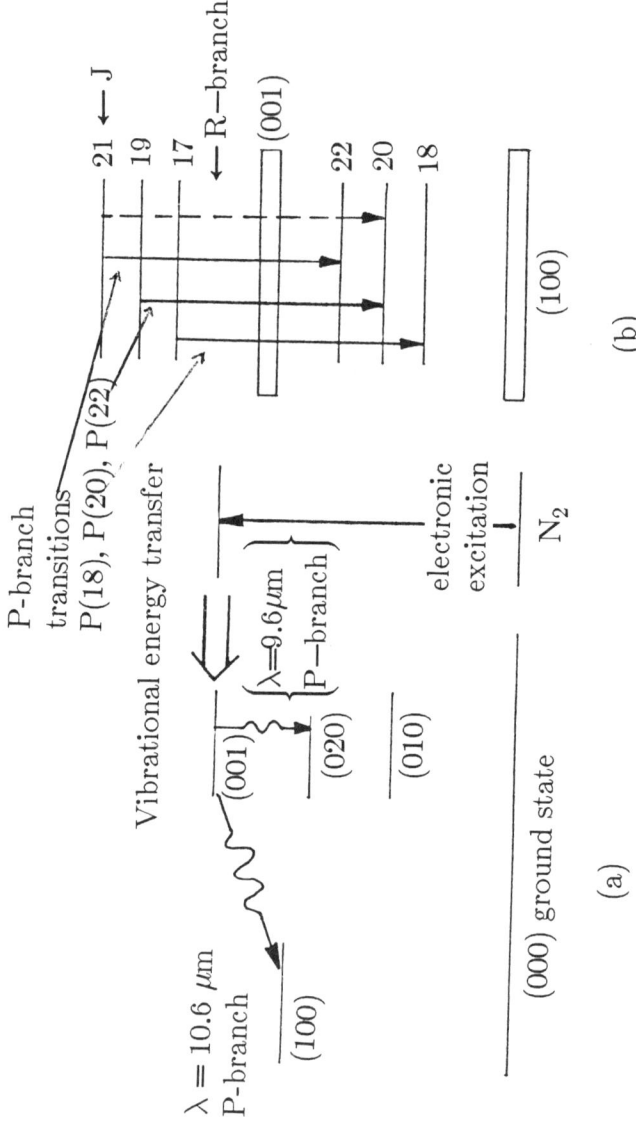

Figure 10.2 (a) CO_2 vibrational energy level diagram (b) some vibration-rotation levels of a vibrating CO_2 molecule. The scale is expanded for (100) and (001), and only a few J-values are shown for clarity.

are generated by using higher input power than the laser can handle continuously, but when the gas temperature passes 150°C the effect terminates, even though the current may remain high. The average power is still constrained by the heat flow rate, just like the CW power.

In many applications beam quality may be more important than beam power. Low-order cavity modes can lead to good beam quality which, in turn, means smaller spot size, requiring less power for equal cutting speeds. The beam symmetry (e.q. circular or elliptical) affects choice of beam motion for a given cutting operation.

Located between each vibrational level are the rotational states designated by the rotational quantum number J, and some J-values are shown in (b) for the vibrational states (001) and (100). The lasing transitions which produces electromagnetic radiation at 10.6 μm (nominal) are *vibration-rotation* transitions between these vibrational levels. There are also transitions which initiate from any of several rotation levels because of the selection rule $\Delta J = \pm 1$. Three such transitions are shown in (b) and the wavelength of these transitions are P(18) - 10.57 μm, P(20) - 10.59 μm, P(22) - 10.61 μm, and P(24) - 10.63 μm (strongest). Note that spectroscopic absorption terminology is used, i.e., a lasing transition labeled "P" is an emission of infrared energy and $\Delta J = +1$ according to (b), but spectroscopic terminology is retained because in *absorption* that same transition corresponds to $\Delta J = -1$, the P-branch of vibration-rotation transition.

The intensities of the absorption lines in the P and R branches first increase with increasing J for both branches but then fall rapidly to zero. Because of the factors which multiply the Boltzmann factor, $e^{-E/kT}$, the absorption intensities of the corresponding lines in the branches are not the same because the frequencies differ in the two branches and because the initial states are different.

Population inversion is necessary for laser action and when the inversion in the vibrational level occurs because of a gas discharge in the tube, the P-branch transitions have more gain than the R-branch transitions, and CO_2 lases only on the P-branch transitions unless special precautions are taken.

Population inversion of the vibrational levels is achieved by an electric discharge through a gaseous mixture of carbon dioxide, nitrogen, and helium. The gas discharge vibrationally excites about 20% of the N_2 molecules to an energy state of N_2 which is almost equal to the (001) state in CO_2. (The actual energy difference is about 0.002 eV). The resonance collision transfer of vibrational energy occurs between the two states. There are no N_2 resonances with the lower laser levels, but when the laser is operative the (100) level and its chain become heavily populated. This chain does not radiate to the ground state but radiates weakly to the lower bending modes. There is a Fermi resonance between (100) and (020) levels and molecular collisions cause these transitions.

Helium is added to increase the power output of the laser by several mechanisms which will not be discussed here. The important lifetimes of the levels are determined by the collision processes. The radiative lifetimes of the states vary from a few milliseconds to a few seconds and average times between molecular collisions is 10^{-5} to 10^{-7} sec.

2 WATER VAPOR LASER

The water vapor laser is a useful source of coherent radiation in the far-infrared region of the electromagnetic spectrum, and it is a useful model for other far-infrared molecular lasers that lase on pure-rotation transitions.

HCN and DCN lasers generate milliwatt CW power on each of several lines, and higher power on many pulsed lines, but only at wavelengths above 70 μm (below 140 cm^{-1}). However, the water vapor laser in pulsed operation oscillates on about a hundred different lines in the range from 20 to 220 μm (500-45 cm^{-1}). Typical peak output powers are 1-100 μW with higher powers available on a few lines when the output coupling is optimized. The gains are fairly low for most of the lines, so the typical water vapor laser is 2-4 meters long. Except for its physical size the water vapor laser is probably one of the simplest gas lasers known.

The transitions responsible for laser action in a water vapor discharge arise from the rotational and vibrational state energy states of an asymmetric top molecule. The gas discharge creates a population inversion between the low-lying stretching mode states and the low-lying bending mode states. The vibrational transition probabilities are normally too small to give sufficient gain for laser action. However, where there occurs a perturbation between two vibrational states, with its consequent mixing of the vibrational wave functions, the vibrational transitions connected with the perturbed levels "steal" some of the rotational transition strength, increasing their gains to the point where laser action may occur. Pure rotational transitions connected with a perturbed level may acquire inversion by virtue of population inversion of the vibrational transitions, so laser action may also occur for these transitions.

At a gas temperature of about 600 K, over 99 percent of the molecules are in the ground vibrational state and rotational levels with $J \leq 13$. The routes for populating the upper levels are either direct excitation from the ground state by electron impact, or excitation of higher-lying vibrational states, which then decay into the states involved in the laser action.

The use of a frequency-selective resonator results in enhanced stable power outputs and new lines that are due to the removal of competitive interactions. This competition between lines is typical of water vapor laser operation and the competing lines are vibrational-rotational transitions between two close lying vibrational states in H_2O.

A striking feature of the emission from an H_2O laser is the lack of apparent regularity in the frequencies of the observed laser lines, whereas in molecular lasers like CO_2 one observes a regular series of lines resulting from various rotational components of the same vibrational transition.

However, the rotational energy states of the water molecule are considerably more complicated than that of a linear molecule like CO_2. The OH radical is known to be an abundant constituent of water vapor discharges; by its recombination one expects some formation of H_2O in rotationally and vibrationally excited states, particularly on the asymmetric stretching mode.

Excitation to higher vibrational states by means of the gas discharge and the recombination of the OH radical favor excitation of the stretching modes, but because of anharmonicity the higher vibrational states become mixtures of pure bending and stretching modes. This mixing, together with the small energy differences involved, increases the cross relaxation among the vibrational states, which tends to equalize the populations in the higher levels. For the lower vibration levels (below about 8000 cm^{-1}) the gaps between the vibrational states become wider, and the rate of relaxation of the bending mode is greater than in the stretching modes. A tendency to produce an inversion in this direction is thus established for an excitation mechanism.

The laser action in a pulsed water vapor laser could be enhanced by the addition of Helium. This enhancement in pulsed operation is a property of a number of gases—Hydrogen, Deuterium, and the noble gases like Helium, Neon and Argon. Helium produces the largest enhancement as well as the most uniform operation over a wide range of pulse repetition rates and pressures.

REFERENCES

1. The Water Vapor Laser, William S. Benedict, Martin A. Pollack, IEEE Journal of Quantum Electronics, QE-5 (1969), pp. 108-124.
2. Enhancement of Laser Action in H_2O by the addition of Helium—W. J. Sarjeant, Eric Brannen—IEEE Journal of Quantum Electronics, QE-5 (1969(, pages 620-621.
3. A Pulsed Brewster Window Water Vapor Laser Operating Between 20 and 120 μ, C. J. Johnson—IEEE Journal of Quantum Electronics, QE-4 (1968), pp. 701-703.
4. Enhancement of Laser Action in Water Vapor by the Addition of Foreign Gases, W. J. Sarjeant, Zdenek Kucerovsky, Eric Brannen, IEEE Journal of Quantum Electronics, QE-6, (1970), pp. 270-271.
5. D. P. Akilt, W. Q. Jeffers, and P. D. Coleman, "Water Vapor Gas Laser Operating at 118-Microns Wavelength," Proceedings of IEEE, Vol. 54, No. 4, pp. 547-551.
6. J. C. Massler, G. Mubner, and P. D. Coleman, "Excitation Mechanism of the Far-infrared Sulfur-Dioxide Molecular Laser," J. Appl. Physics, Vol 44, No. 2, pp. 795-801.
7. Davis, J. C., Advanced Physical Chemistry, Ronald Press, 1965.
8. Wells, A. F., Structural Inorganic Chemistry, Oxford Press, 1962.

INDEX

absorption 2,21,23,25,27
ammonia, 63
anharmonicity, 31,35
anti-Stokes line, 53,56
asymmetric top, 17

band center, 45
band head, 48
bandwidth, 41
Boltzmann, 26

$ClCH_3$, 67
CO_2 laser, 118
centrifugal stretching, 28
character, 78,88,91,94,95
character table, 83
class, 66
combination relation, 36,48
combinations, selection rule, 100
conjugate element, 66

coordinate system, 67
correspondence principle, 22
covering operation, 62

dispersive, 5
double degeneracy, 102

eigenfunction, 14
eigenvalue, 14
electric susceptibility, 112
emission, 2,21
ethylene molecule, 72
expansion theorem, 88

far infrared, 3
Fermi resonance, 37
filter, 23
Fortrat diagram, 48
Fourier integral, 41

125

Fourier transform, 6
frequency, 22
fundamental, 45

grating, 115
group, 60

H_2O laser, 122
Hamiltonian, 13
harmonic oscillator, 32
harmonics, 35
heteronuclear, 33
Hooke's Law, 31
hot band, 37

improper rotation, 67
infrared, 3
infrared selection rules, 90
inversion, 111
isomorphism, 65

J quantum number, 25
J-value, 25

k-value, 89

laser, 109, 117
linewidth, 113
Lorentz lineshape, 25

microwave, 3
mode, 115
moment of inertia, 18
multiplication table, 64

Newton's law, 32

normalization, 12

operator, 13, 60,
optical path, 115
order, 65
overtones, 35, 105
overtones, selection rules, 102

P-branch, 45
particle, 9
period, 65
perturbation, 35
photon, 6
Planck's constant, 6
polarizability, 52, 91
polarization, 52
potential, 32, 35
precession, 25
principle axis, 7
probability, 12
pump, 111

quality factor, 40, 49
quantum mechanics, 9

Raman effect, 33, 52, 55
Raman selection rules, 93
R-branch, 45
Rayleigh, 53
reduced mass, 19
reducible, 81
relative intensity, 26
representation, 78
resonance, 39
rigid rotor, 18
rotation of axes, 76

S_2Cl_2, 69
scattering, 2

Schrodinger, 10
selection rules, 14, 25
species, 85
spectra, 2
spectrograph, 2
spectrometer, 2
spectroscopy, 2
spherical top, 17
stationary state, 14
step, 80
stimulated emission, 110
Stokes line, 53,56
subgroup, 65
symmetric top, 17

term value, 21,25,34
triple degeneracy, 102

unitary matrix, 80
unitary transformation, 80

vibration, 31,43
visible, 3

water molecule, 70
wave, 11
wave function 11,12,20
wave number, 6

zero order, 44

MAY 0 2 1990